Lecture Notes in Artificial Intelligence 7607

Subseries of Lecture Notes in Computer Science

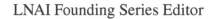

Longbing Cao Yifeng Zeng
Andreas L. Symeonidis Vladimir I. Gorodetsky
Philip S. Yu Munindar P. Singh (Eds.)

Agents and Data Mining Interaction

8th International Workshop, ADMI 2012
Valencia, Spain, June 4-5, 2012
Revised Selected Papers

Springer

Series Editors

Randy Goebel, University of Alberta, Edmonton, Canada
Jörg Siekmann, University of Saarland, Saarbrücken, Germany
Wolfgang Wahlster, DFKI and University of Saarland, Saarbrücken, Germany

Volume Editors

Longbing Cao, University of Technology, Sydney, NSW, Australia
E-mail: longbing.cao@uts.edu.au

Yifeng Zeng, Teesside University, Middlesbrough, UK
E-mail: y.zeng@tees.ac.uk

Andreas L. Symeonidis
Aristotle University, Thessaloniki, Greece
E-mail: asymeon@issel.ee.auth.gr

Vladimir I. Gorodetsky
Institute for Informatics and Automation, St. Petersburg, Russia
E-mail: gor@mail.iias.spb.su

Philip S. Yu, University of Illinois, Chicago, IL, USA
E-mail: psyu@cs.uic.edu

Munindar P. Singh, North Carolina State University, Raleigh, NC, USA
E-mail: mpsingh@ncsu.edu

ISSN 0302-9743 e-ISSN 1611-3349
ISBN 978-3-642-36287-3 e-ISBN 978-3-642-36288-0
DOI 10.1007/978-3-642-36288-0
Springer Heidelberg Dordrecht London New York

Library of Congress Control Number: 2012956025

CR Subject Classification (1998): I.2.11, H.2.8, I.2.6, K.4.4, H.3.4, H.5.3, H.4.3, J.2

LNCS Sublibrary: SL 7 – Artificial Intelligence

Typesetting: Camera-ready by author, data conversion by Scientific Publishing Services, Chennai, India

Printed on acid-free paper

Springer is part of Springer Science+Business Media (www.springer.com)

Preface

We are pleased to present the proceedings of the 2012 International Workshop on Agents and Data Mining Interaction (ADMI-12), which was held jointly with AAMAS 2012.

In recent years, agents and data mining interaction (ADMI, or agent mining) has emerged as a very promising research field. Following the success of ADMI-06 in Hongkong, ADMI-07 in San Jose, AIS-ADM-07 in St Petersburg, ADMI-08 in Sydney, ADMI-09 in Budapest, ADMI-10 in Toronto, and ADMI-11 in Taipei, ADMI-12 provided a premier forum for sharing research and engineering results as well as potential challenges and prospects encountered in the coupling between agents and data mining.

The ADMI-12 workshop encouraged and promoted theoretical and applied research and development that aims to:

- Exploit agent-enriched data mining and demonstrate how intelligent agent technology can contribute to critical data-mining problems in theory and practice
- Improve data mining-driven agents and show how data mining can strengthen agent intelligence in research and practical applications
- Explore the integration of agents and data mining toward a super-intelligent system
- Discuss existing results, new problems, challenges, and impacts of the integration of agent and data-mining technologies as applied to highly distributed heterogeneous systems, including mobile systems operating in ubiquitous and P2P environments; and
- Identify challenges and directions for future research and development on the synergy between agents and data mining.

The 16 papers accepted by ADMI-12 are from ten countries. ADMI-12 submissions cover areas from North America, Europe to Asia, indicating the booming of agent-mining research globally. The workshop also included two invited talks by distinguished researchers Victor Lesser and Wolfgang Ketter.

Following ADMI-09, the papers accepted by ADMI-12 have been revised and published as an LNAI proceedings volume by Springer. We appreciate Springer, in particular Alfred Hofmann, for the continuing publication support.

ADMI-12 was sponsored by the Special Interest Group: Agent-Mining Interaction and Integration (AMII-SIG: www.agentmining.org). We appreciate the guidance of the Steering Committee.

More information about ADMI-12 is available from the workshop website: http://admi12.agentmining.org/.

Finally, we appreciate the contributions made by all authors, Program Committee members, invited speakers, panelists, and the AAMAS 2012 workshop local organizers.

June 2012

Philip S. Yu
Munindar P. Singh
Longbing Cao
Yifeng Zeng
Andreas L. Symeonidis
Vladimir Gorodetsky

Organization

General Chairs

Philip S. Yu	University of Illinois at Chicago, USA
Munindar P. Singh	North Carolina State University, USA

Workshop Co-chairs

Longbing Cao	University of Technology Sydney, Australia
Yifeng Zeng	Teesside University, UK
Andreas L. Symeonidis	Aristotle University of Thessaloniki, Greece
Vladimir Gorodetsky	Russian Academy of Sciences, Russia

Workshop Organizing Co-chair

Xinhua Zhu	University of Technology Sydney, Australia
Hua Mao	Aalborg University, Denmark

Program Committee

Ahmed Hambaba	San Jose State University, USA
Ajith Abraham	Norwegian University of Science and Technology, Norway
Andrea G.B. Tettamanzi	University of Milan, Italy
Andreas L. Symeonidis	Aristotle University of Thessaloniki, Greece
Andrzej Skowron	Institute of Decision Process Support, Poland
Bo An	University of Southern California, USA
Bo Zhang	Tsinghua University, China
Daniel Kudenko	University of York, UK
Daniel Zeng	Arizona University, USA
David Taniar	Monash University, Australia
Deborah Richards	Macquarie University, Australia
Dionysis Kehagias	Informatics and Telematics Institute, Greece
Eduardo Alonso	University of York, UK
Eugenio Oliveira	University of Porto, Portugal
Frans Oliehoek	Massachusetts Institute of Technology, USA
Gao Cong	Nanyang Technological University, Singapore
Henry Hexmoor	University of Arkansas, USA

Ioannis Athanasiadis	Democritus University of Thrace, Greece
Jason Jung	Yeungnam University, Korea
Joerg Mueller	Technische University Clausthal, German
Juan Carlos Cubero	University de Granada, Spain
Katia Sycara	Carnegie Mellon University, USA
Kazuhiro Kuwabara	Ritsumeikan University, Japan
Kim-leng Poh	National University of Singapore
Leonid Perlovsky	AFRL/IFGA, USA
Luis Otavio Alvares	Universidade Federal do Rio Grande do Sul, Brazil
Martin Purvis	University of Otago, New Zealand
Michal Pechoucek	Czech Technical University, Czech Republic
Mingyu Guo	University of Liverpool, UK
Nathan Griffiths	University of Warwick, UK
Pericles A. Mitkas	Aristotle University of Thessaloniki, Greece
Ran Wolff	Haifa University, Israel
Seunghyun Im	University of Pittsburgh at Johnstown, USA
Stefan Witwicki	Instituto Superior Técnico, Portugal
Sung-Bae Cho	Yonsei University, Korea
Sviatoslav Braynov	University of Illinois at Springfield, USA
Tapio Elomaa	Tampere University of Technology, Finland
Valerie Camps	University Paul Sabatier, France
Vladimir Gorodetsky	SPIIRAS, Russia
Wen-Ran Zhang	Georgia Southern University, USA
William Cheung	Hong Kong Baptist University, HK
Xudong Luo	Sun Yat-sen University, China
Yan Wang	Macquarie University, Australia
Yinghui Pan	Jiangxi University of Finance and Economics, China
Yingke Chen	Aalborg University, Denmark
Yves Demazeau	CNRS, France
Zbigniew Ras	University of North Carolina, USA
Zili Zhang	Deakin University, Australia
Zinovi Rabinovich	University of Southampton, UK

Steering Committee

Longbing Cao	University of Technology Sydney, Australia (Coordinator)
Edmund H. Durfee	University of Michigan, USA
Vladimir Gorodetsky	St. Petersburg Institute for Informatics and Automation, Russia
Hillol Kargupta	University of Maryland Baltimore County, USA

Matthias Klusch	DFKI, Germany
Michael Luck	King's College London, UK
Jiming Liu	Hong Kong Baptist University, China
Pericles A. Mitkas	Aristotle University of Thessaloniki, Greece
Joerg Mueller	Clausthal University of Technology, Germany
Ngoc Thanh Nguyen	Wroclaw University of Technology, Poland
Carles Sierra	Artificial Intelligence Research Institute of the Spanish Research Council, Spain
Andreas L. Symeonidis	Aristotle University of Thessaloniki, Greece
Gerhard Weiss	University of Maastricht, The Netherlands
Xindong Wu	University of Vermont, USA
Philip S. Yu	University of Illinois at Chicago, USA
Chengqi Zhang	University of Technology Sydney, Australia

Table of Contents

Part I: Invited Talks

Part II: Agents for Data Mining

Part III: Data Mining for Agents

Part IV: Agent Mining Applications

Part I

Invited Talks

Organizational Control for Data Mining with Large Numbers of Agents

Victor Lesser

University of Massachusetts Amherst, USA
lesser@cs.umass.edu

Abstract. Over the last few years, my research group has begun exploring the issues involved in learning when there are hundreds to thousands of agents. We have been using the idea of organization control as a low overhead way of coordinating the learning of such large agent collectives. In this lecture, the results of this research will be discussed and its relationship to issues in distributed data mining.

L. Cao et al.: ADMI 2012, LNAI 7607, p. 3, 2013.

Competitive Benchmarking: Lessons Learned from the Trading Agent Competition

Wolfgang Ketter

Erasmus University Rotterdam, The Netherlands
wketter@rsm.nl

Abstract. Many important developments in artificial intelligence have been stimulated by organized competitions that tackle interesting, difficult challenge problems, such as chess, robot soccer, poker, robot navigation, stock trading, and others. Economics and artificial intelligence share a strong focus on rational behavior. Yet the real-time demands of many domains do not lend hemselves to traditional assumptions of rationality. This is the case in many trading environments, where self-interested entities need to operate subject to limited time and information. With the web mediating an ever broader range of transactions and opening the door for participants to concurrently trade across multiple markets, there is a growing need for technologies that empower participants to rapidly evaluate very large numbers of alternatives in the face of constantly changing market conditions. AI and machine-learning techniques, including neural networks and genetic algorithms, are already routinely used in support of automated trading scenarios. Yet, the deployment of these technologies remains limited, and their proprietary nature precludes the type of open benchmarking that is critical for further scientific progress.

The Trading Agent Competition was conceived to provide a platform for study of agent behavior in competitive economic environments. Research teams from around the world develop agents for these environments. During annual competitions, they are tested against each other in simulated market environments. Results can be mined for information on agent behaviors, and their effects on agent performance, market conditions, and the performance and behavior of competing agents. After each competition, competing agents are made available for offline research. We will discuss results from various competitions (Travel, Supply-Chain Management, Market Design, Sponsored Search, and Power Markets).

L. Cao et al.: ADMI 2012, LNAI 7607, p. 4, 2013.

Part II

Agents for Data Mining

Supporting Agent-Oriented Software Engineering for Data Mining Enhanced Agent Development

Andreas L. Symeonidis[1,2], Panagiotis Toulis[3], and Pericles A. Mitkas[1,2]

[1] Electrical & Computer Engineering Department,
Aristotle University of Thessaloniki
[2] Informatics and Telematics Institute, CERTH
Thessaloniki, Greece
asymeon@eng.auth.gr, mitkas@auth.gr
[3] Department of Statistics, Harvard University, Boston, MA, USA
ptoulis@fas.harvard.edu

Abstract. The emergence of Multi-Agent systems as a software paradigm that most suitably fits all types of problems and architectures is already experiencing significant revisions. A more consistent approach on agent programming, and the adoption of Software Engineering standards has indicated the pros and cons of Agent Technology and has limited the scope of the, once considered, programming 'panacea'. Nowadays, the most active area of agent development is by far that of intelligent agent systems, where learning, adaptation, and knowledge extraction are at the core of the related research effort. Discussing knowledge extraction, data mining, once infamous for its application on bank processing and intelligence agencies, has become an unmatched enabling technology for intelligent systems. Naturally enough, a fruitful synergy of the aforementioned technologies has already been proposed that would combine the benefits of both worlds and would offer computer scientists with new tools in their effort to build more sophisticated software systems. Current work discusses *Agent Academy*, an agent toolkit that supports: a) rapid agent application development and, b) dynamic incorporation of knowledge extracted by the use of data mining techniques into agent behaviors in an as much untroubled manner as possible.

1 Introduction

More than a decade ago, agents appeared as a 'hype' that was abstract enough to fit any given or future problem. It is only recently that their range of applicability has been narrowed down to specific application domains (e.g. Grid computing [7], electronic auctions [8], autonomic computing [11] and social networks [9]) that exploit the beneficial characteristics of agents. Still, software practitioners are reluctant in incorporating agent solutions to solve real-world problems, even in the case of the above-mentioned domains, where agents have proven to be efficient. This reluctance has been attributed to many reasons that range from

L. Cao et al.: ADMI 2012, LNAI 7607, pp. 7–21, 2013.

the lack of consensus in definitions and the interdisciplinary nature of agent computing to the lack of tools and technologies that truly allege the real benefits of *Agent Technology* (AT) [12]. In fact, AT still seems like it is missing its true scope.

May one take a closer look at the domains that AT is considered as the proper programming 'metaphor', one may notice that, apart from their dynamic nature and their versatility, these domains have another thing in common: they generate vast quantities of data at extreme rates. Data from heterogeneous sources, of varying context and of different semantics become available, dictating the exploitation of this "pile" of information, in order to become useful bits and pieces of knowledge, the so-called *knowledge nuggets* [13]. Nuggets that will be used by software systems and will add intelligence, in the sense of adaptability, trend identification and prevention of decision deadlocks.

To this end, *Data Mining* (DM) appears to be a suitable paradigm for extracting useful knowledge. The application domain of Data Mining and its related techniques and technologies have been greatly expanded in the last few years. The development of automated data collection tools has fueled the imperative need for better interpretation and exploitation of massive data volumes. The continuous improvement of hardware, along with the existence of supporting algorithms has enabled the development and flourishing of sophisticated DM methodologies. Issues concerning data normalization, algorithm complexity and scalability, result validation and comprehension have been successfully dealt with [1,10,15]. Numerous approaches have been adopted for the realization of autonomous and versatile DM tools to support all the appropriate pre- and post-processing steps of the knowledge discovery process in databases [5,6].

From all the above, the synergy of AT with DM seems promising towards providing a thrust in the development and establishment of intelligent agent systems [3,4]. Knowledge hidden in voluminous data repositories can be extracted by data mining, and provide the inference mechanisms or simply the behavior of agents and multi-agent systems. In other words, these knowledge nuggets may constitute the building blocks of agent intelligence. We argue that the two, otherwise diverse, technologies of data mining and intelligent agents can complement and benefit from each other, yielding more efficient solutions.

However, while the pieces are already there, the puzzle is far from complete. No existing tool provides solutions to the theoretical problems of researchers striving to seamlessly integrate data mining with agent technology, nor to the practical issues developers face when attempting to build even simple multi-agent systems (MAS).

Agent Academy (AA) is a practical approach to the problem of establishing a working synergy between software agents and data mining. As a lower CASE (Computer Aided Software Engineering) tool, Agent Academy combines the power of two proven and robust software packages (APIs), namely JADE [2] and WEKA [15], and along with the related theoretical framework, AA sketches a new methodology for building data mining-enabled agents. As a development tool, Agent Academy is an open source IDE with features like structured code editing, graphical debugging and beginner-friendly interface to data mining procedures. Finally, as a framework, Agent Academy is a complete solution enabling the creation of software agents at any level of granularity, with specific interest on intelligent systems and decision making.

Technically, Agent Academy acts as a hub between the agent development phase and the data mining process by adding models as separate and independent threads into the agent's thread pool. These execution units, usually called behaviors in software agent terminology, are reprogrammable, reusable and even movable among agents. Using Agent Academy, intelligent agents are not an afterthought, but rather the very basic capability that this engineering tool has to offer.

The remainder of this chapter is organized as follows: Section 2 presents the methodology to follow in order to ensure the synergy between AT and DM. Section 3 discusses Agent Academy in detail, the development framework for DM-enriched MAS, while Section 4 summarizes work presented and discusses future extensions.

2 Integrating Agents and Data Mining

The need to couple AT with DM comes from the emergent need to improve agent systems with knowledge derived from DM, so as to strengthen their existence in the software programming scenery. Nevertheless, coupling of the two technologies does not come seamlessly, since the inductive nature of data mining imposes logic limitations and hinders the application of the extracted knowledge on deductive systems, such as multi-agent systems. The methodology and the supporting tool described within the context of this paper take all the relevant limitations and considerations into account and provide a pathway for employing data mining techniques in order to augment agent intelligence.

Knowledge extraction capabilities must be present in agent design, as early as in the agent modeling phase. During this process, the extracted models of the DM techniques applied become part of the knowledge model of agents, providing them with the ability to enjoy DM advantages.

In Symeonidis and Mitkas [14], the reader may find more details on the unified methodology for transferring DM extracted knowledge into newly created agents. Knowledge diffusion is instantiated in three different ways, always with respect with the levels of diffusion of the extracted knowledge (DM on the application level of a MAS, DM on the behavioral level of a MAS, DM on the evolutionary agent communities). The iterative process of retraining through DM on newly acquired data is employed, in order to enhance the efficiency of intelligent agent behavior.

As already mentioned, the methodology presented is also supported by the respective toolkit. Using Agent Academy, the developer may automate (or semi-automate) several of the tasks involved in the development and instantiation of a MAS. He/she follows the steps shown in Figure 1, in order to build a DM-enhanced MAS in an untroubled manner. Details on the Agent Academy framework as discussed next.

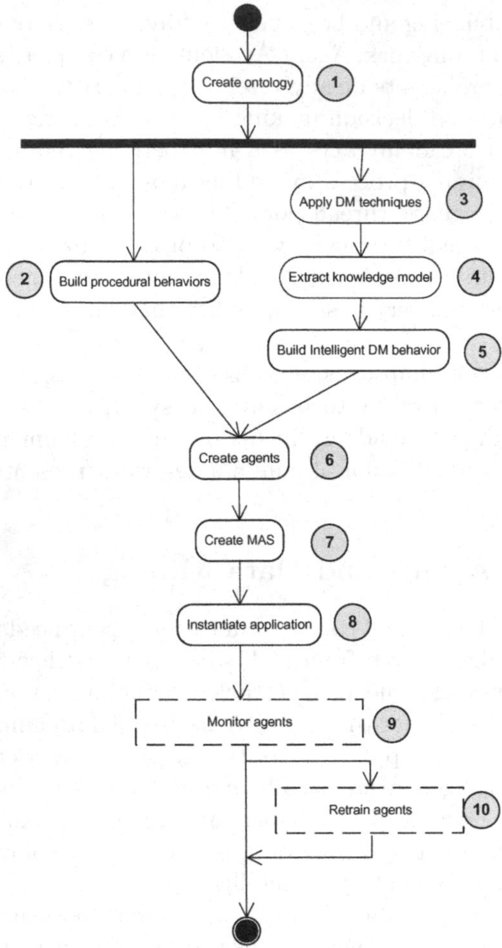

Fig. 1. The MAS development steps

3 Agent Academy

3.1 Introduction

Agent Academy is an open-source framework and integrated development environment (IDE) for creating software agents and multi-agent systems, and for augmenting agent intelligence through data mining. It follows the aforementioned methodology, in order to support the seamless integration of Agent Technology and Data Mining.

The core objectives, among others, of Agent Academy are to:

- Provide an easy-to-use tool for building agents, multi-agent systems and agent communities
- Exploit Data Mining techniques for dynamically improving the behavior of agents and the decision-making process in multi-agent systems

- Serve as a benchmark for the systematic study of agent intelligence generated by training them on available information and retraining them whenever needed.
- Empower enterprize agent solutions, by improving the quality of provided services.

Agent Academy has been implemented upon the JADE and WEKA APIs (Application Programming Interfaces), in order to provide the functionality it promises. The initial implementation of AA was funded under the fifth Framework Program, where the theoretical background was formulated. After the successful completion of the project the second version of Agent Academy (AA-II) was built, where emphasis was given on the user interface and functionality. Agent Academy is built in Java and is available at Sourceforge (http://sourceforge.net/projects/agentacademy). The current release contains 237 Java source files and is spun on over 28,000 lines of code.

The original Agent Academy vision stands on the edge of being an intelligent agent research tool. Nevertheless, considerable effort was given in order to provide adequate quality level and user-friendliness, so as to support industrial-scope agent applications. To this end, one of the pivotal requirements for AA was to provide maximum functionality with the minimum learning curve. Agent Academy has been built around well-known concepts and practices, not trying to introduce new APIs and standards, rather to increase applicability of the already established ones. The novelty of AA lies on the fruitful integration of agent and data mining technologies, and the methodology for doing so. In fact, Agent Academy proposes a new line of actions for creating DM-enhanced agents, through a simple and well-defined workflow. Prior to unfolding the details of this workflow we define the concepts and conventions AA is founded on (Table 1).

In a typical scenario the developer should follow the methodology described in the previous section in order to build a new agent application. He/she should first create behaviors, as orthogonal as possible. while in a parallel process, he/she should build the data mining models which apply to the application under development. Next, everything should be organized/assigned to agents, in order to finally build the multi-agent system by connecting the developed software agents.

3.2 Agent Academy Architecture

The AA IDE offers a complete set of modules that can guide and help through the entire process. These modules, namely the *Behavior Design Tool*, the *Agent Design Tool*, and the *MAS Design Tool* support the respective development phases of an Agent Academy project through graphical user interfaces, while they also support code editing, debugging and compilation.

Nevertheless, AA follows a component-based approach, where each of the modules is loosely coupled with the others. This way the user may decide to build (JADE compliant) behaviors or (WEKA) data mining models with any other tool/framework and just import them in AA, just by providing the relative path of their location. In the following sections we first provide a qualitative description of the three core modules, and then continue with a more technical description on them.

Table 1. Agent Academy conventions

Behaviors	Blocks of code sharing the same functional description, which can, conceptually, be grouped together. Based on the JADE paradigm, behaviors are essentially distinct threads of execution that run in parallel and are assigned to agents, in order to exhibit desired (agent) properties
Intelligent DM behaviors	Technically, the combination of plain (procedural) agent behaviors with data mining models. Any possible type of data mining generated knowledge can serve as the element of an intelligent behavior, e.g. classification intelligent behavior, clustering intelligent behavior, etc. Such types of behaviors are based on data mining models built by the use of the WEKA API (most of the times offline). AA then uptakes the task of transforming the WEKA-generated model into a compiled JADE behavior which can, after that point, operate indistinguishably in any JADE-developed MAS. This type of behavior is a special case of an Intelligent behavior
Agents	Distinct software entities, programming metaphors, embracing a groups of behaviors. Agents in AA can communicate with others through well established protocols (the FIPA Agent Communication Language - ACL)
Intelligent DM Agents	Agents that have at least one intelligent DM behavior in their collection of behaviors. In practice, such agents can emulate any kind of high-level cognitive task backed by the corresponding DM model (classify, cluster, associate etc). This type of agents is a special case of an intelligent agent
Multi-agent Systems	Collections of agents, intelligent or not, operating under the same environment

Agent Academy Overview. Agent Academy organizes work into projects. Each project has a private space that contains all the necessary elements code for the development and instantiation of an intelligent MAS. These elements may be of two types: i) software snippets, i.e. programming code resulting to behavior classes and agent classes, and ii) data (either in the form of a file or as database connection), where data mining will be applied on in order to generate DM models.

Creating Agent Behaviors. The notion of agent behavior is very popular among agent-oriented software methodologies and represents essentially a block of code that encapsulates an execution thread, aiming to fulfill one, or a few well-defined tasks. Specifically in the JADE API, which is employed by Agent Academy, an agent behavior is realized as a distinct Java class.

The *Behavior Design Tool* (BDT) of Agent Academy implements a minimal code editing tool with code automation functionality. BDT comprises two panes: the BDT toolbar and the code editing area. Although minimal, BDT should be considered as a fully functional agent behavior creation tool, since it provides features such as automatic code generation for typical agent code blocks (such sending/receiving FIPA ACL messages), structured code tools, text editing and compilation options. Apart from plain (procedural) behaviors, Agent Academy defines another category of agent behaviors, namely the *Intelligent DM Behaviors*, which encapsulate DM models in an AA-compliant code wrapper. These "intelligent" pieces of code are created through a 5-step process, which employs the WEKA API to build the data mining models and then compile them into JADE-compliant behaviors.

Creating Software Agents. The *Agent Design Tool* (ADT) is the functional equivalent of BDT for agent creation. The definition of "agency" within the development context of Agent Academy is absolutely generic: an agent is simply thought of as a meaningful grouping of agent behaviors, aiming to a specific set of tasks, following a specific workflow. Thus, in order to build an agent in AA, the developer needs only to assign the agent with a set of behaviors and a sequence of behavior execution. All project behaviors, both *plain* and *intelligent*, are available for insertion to the agents' execution pool, through the use of automatic IDE tools.

ADT comprises the code editing area and the ADT toolbar that provides code generation and debugging functions. ADT can be used to assign behaviors to agents, compile the generated code, generate agent class files, and debug agent execution.

Instantiating Multi-agent Systems. The *MAS Design Tool* (MDT) allows the developer to organize agents into multi-agent systems. It provides the tools to initialize, pause and terminate the MAS, as well as to monitor agent state (alive, dead, etc). Typically, this step concludes an Agent Academy project, i.e. the development and instantiation phase of a MAS.

3.3 Technical Details of Agent Academy

Upon project creation, Agent Academy creates a specific folder structure under the directory:

```
<INSTALL_DIR>/user/projects/<PROJECT_NAME>
```

In the previous section we provided an overview of the AA modules. Here, we further elaborate on them.

The Behavior Design Tool. The first step in the Agent Academy workflow is the creation of the agent behaviors which are the software blocks that encapsulate the functionality of the agents. These behaviors are typical Java classes and can be created using the Behavior Design Tool.

BDT provides a set of features that may assist in the creation of the agent behaviors. AA behaviors are extensions of well-defined JADE behaviors; this implies that developers already familiar with the JADE framework may use BDT to build behaviors directly. Even in the case of unexperienced developers, though, AA provides automated code generators to help the user through the behavior creation process. The workflow of creating a new behavior entails the following:

1. Initializing the behavior, setting names and identifiers, and defining the type of behavior
2. Adding fields, methods and blocks of code, through the *Structured Edit Action* - SEA module, a specific-purpose module of the Agent Academy IDE. The developer may, of course, write the communication source code himself/herself, in case he/she prefers
3. Specifying FIPA-compliant agent communication, through the respective messaging buttons located on the BDT toolbar, which automatically generate code blocks for sending and receiving ACL messages. The developer may, of course, write the communication source code himself/herself, in case he/she prefers
4. Optionally documenting the behavior, by providing a small description and saving it to the project behaviors *notepad*
5. Saving the newly created source code and compiling, directly from the AA environment.

This workflow is supported by the functions depicted in Figure 2.

Upon behavior initialization, the developer is requested to define the name of the behavior, the sub-package the behavior should belong and its type. All the fundamental JADE behaviors types are supported within the context of AA:

- `OneShotBehavior`, where a behavior is executed only once
- `CyclicBehavior`, where a behavior is executed ad infinitum
- `TickerBehavior`, where a behavior is executed periodically, with respect to a user-defined time slot
- `WakerBehavior`, where a behavior is executed when certain conditions are met.

In addition, the JADE framework provides the archetypical behavior class (`SimpleBehavior`), which can be modified in order to produce any kind of functionality inside the generated behavior.

As already mentioned, the AA IDE supports automated code authoring for generating FIPA-compliant ACL messages. This kind of functionality, along with the documentation options AA provides, allow for **rapid agent-based application development**.

One of the most interesting features of the BDT is the SEA, which offers a systematic way to create Java class members along with a visual 'summary' of the behavior functionality. Through the SEA feature developers may:

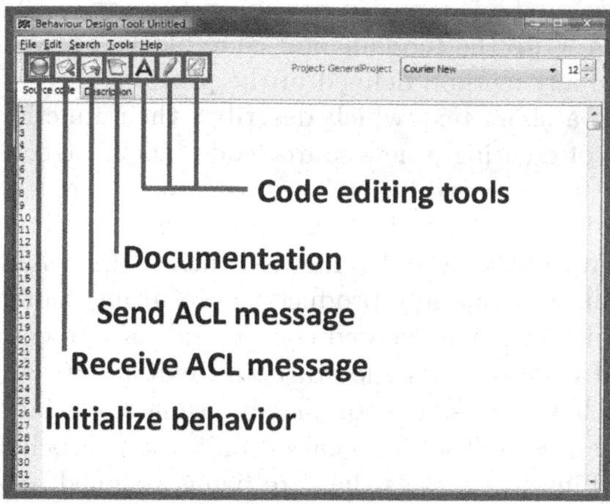

Fig. 2. Functionality provided for creating agent behaviors

- Define class/instance fields and automatically generate `get/set` methods for them (much like the Properties of the C# language)
- Define methods by providing their signatures
- Create documented code blocks.

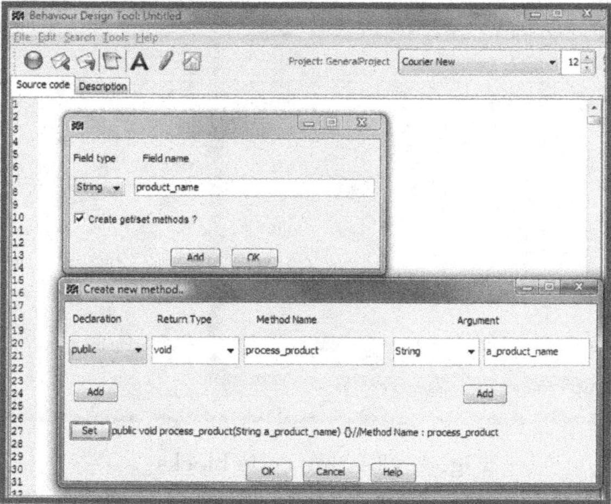

Fig. 3. Creating fields and methods

Developers have the ability to transfer code blocks among methods and also dynamically edit their content. The *Description* tab of the source code editor can be used to provide with a brief summary of what the behavior really does. As long as the developer conforms to the AA code structuring standards, this approach can be quite beneficial for code clarity and maintainability.

Figure 3 illustrates the use of SEA in defining a class field named `product_name` of type `String`, creating the `get/set` methods for it and then defining a new method named `process_product()`. May one use the code editor, one can write the fundamental code blocks of the behavior, which can be embedded to any method defined in the previous steps. Optionally they can be tagged with a short text which describes their functionality. Figure 4 depicts the process of creating a new source code block, tagged as "*PROCESS THE PRODUCT*", and then adding the source code to implement the desired functionality. By using the *Method* process list, one can embed the newly created code block to any of the available methods inside the behavior. In this case Agent Academy will automatically produce the resulting source code, but will also inject some additional AA-oriented code useful for debugging purposes. In fact, the additional code denotes the start and exit points of every behavior created with the Behavior Design Tool, and by using a special debugging component of the Agent Design Tool, it is easy monitor which behavior is executed at any given time. Finally, the tags that are being assigned to each code block are listed in succession at the *Description* tab, which summarizes quickly and efficiently the behavior functionality.

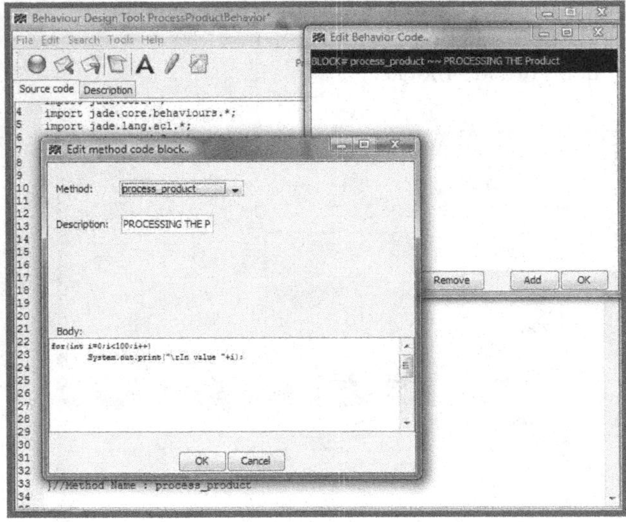

Fig. 4. Creating code blocks

It should be denoted that all source code libraries needed both for performing JADE-related tasks and Agent Academy structures are automatically imported during behavior initialization.

Intelligent Data Mining Behaviors. As already discussed, an Intelligent DM behavior in Agent Academy is a typical JADE behavior that also incorporates knowledge extracted by the application of DM techniques on problem related data. This knowledge is represented by a DM model, like the ones that are generated by the WEKA API. When building such an intelligent behavior, the user typically follows all the steps of the KDD process (section 2). Through the AA-adapted WEKA GUI, he/she supplies the dataset DM will be applied on, performs preprocessing, feature selection, algorithm parametrization, and model evaluation, like in any given DM problem. After the DM model has been built, the developer has to either approve it or request retraining, which implies reconfiguration of the KDD process at any point (preprocessing, algorithm parameters, etc). Upon model approval, the developer calls Agent Academy to embed the generated model in a JADE behavior and compile it.

Following the AA conventions in the development of a AA project, data to be mined are stored in a specific location:

```
<INSTALL_DIR>/user/projects/<PROJECT_NAME>/data/DMRepository/
```

To initiate the DM behavior building process, the user has to navigate from "*Tools*" to "*Create a Data Mining Behavior*", from the initial screen of Agent Academy. In order to wrap the DM model in a JADE-compliant schema, AA creates a JADE `CyclicBehavior`, configured to wait for ACL messages adhering to the following two constraints:

- The Ontology of the message should be set to the constant `AAOntoloty.` *ONTOLOGY_NAME*
- The ACL message protocol should be set according to the specific operation that takes place. For example, if the DM model performs classification, the protocol should be set to `AAOntology.CLASSIFY_REQUEST`

These two constraints guarantee that an agent with the DM behavior will succeed in communicating with other agents on the specific topic (e.g. classification task). In addition, the actual message content should be an object of the specific type class that extends the `org.aa.ontology.`*DM_TECHNIQUE_REQUEST* class (in case of classification this should be the `org.aa.ontology.Classify Request` class). This object, in fact, wraps the actual data tuple the model will be applied on.

In response, the agent executing the intelligent behavior shall generate an ACL message with the protocol field set to `AAOntology.CLASSIFY_RESPONSE` (in the case of classification) and the actual content of the message will be a serialized object, namely of type `org.aa.ontology.ClassifyResponse`. This object, in fact, wraps the output of the DM model, with respect to the input data tuple the sender provided.

The actual mechanism of building an intelligent DM behavior is a bit more complicated. When the developer builds such a behavior after having produced the respective DM model, AA produces three additional files with extensions *.aadm*, *.aainst* and *.dat*. These files contain the training dataset the model was built on, along with the generated model itself. When an operation, such as a classification,, is requested, Agent Academy retrieves the "*.aainst*" file and regenerates it as a functional Java class, which is able to carry out the data mining operation. This way, both the model and the agent behavior can be created on-the-fly. The requests and responses that agents exchange in this context should be subclasses of the *Request* and *Response* classes defined in the `org.aa.ontology` package. AA by default defines such classes, like `ClassifyRequest` and `ClusterRequest`, nevertheless the user is allowed to define more complex structures. These objects practically carry the data for processing in a field of type `org.aa.tools.DMList`, which is essentially a vector of values, *double* or *String* type.

3.4 Agent Design Tool

Agent Academy treats agents as logical groupings of software behaviors, based on the idea that an agent is completely defined by the way it actually acts. Thus, ADT provides all the functionality for inserting and organizing behaviors in an agent execution pool. Figure 5 illustrates the features of the Agent Design Tool, listed below:

1. **Agent definition**, during which the user sets the name of the agent and the package to which it belongs
2. **Behavior assignment**, which browses for project behaviors and lists them for the user to assign. This feature is available through the "*Add Behavior*" option.
3. **Text editing and compilation options**, where the user may manually edit the source code, in case he/she deems necessary. Through this option the developer may also invoke the Java compiler, so that the new agent class is generated and is mapped to the agent
4. **Agent Monitoring**, which provides a graphical tool for validating agent execution
5. **Intelligent DM Behavior assignment**, which enables the user to assign intelligent DM behaviors to the agent behavior pool. Through this feature, the user may also request for model retraining of a specific DM Behavior, in case of poor performance or existence of update data. Retraining may be performed on-the-fly, since all the model parameters are available and are loaded on behavior initialization at runtime.

In order to initialize an agent, the developer has to select a name and optionally select the package to which it shall be assigned to. Agent Academy considers a default package for every behavior or agent, and any user-defined package should be a sub-package of the default one. Thus, if `<PROJECT_NAME>` denotes the project name, behaviors are compiled by default as classes and are stored in the `<PROJECT_NAME>.behaviors` package, while agents are compiled and stored in the `<PROJECT_NAME>.agents` package. Naturally, the user is allowed to organize class files into more packages, nevertheless it is mandatory all of them to be sub-packages of the default packages. When the user browses for behaviors to add to the agent, a list of Java classes are available through a regular 'Open File Dialog'. After the behaviors have been added, Agent Academy automatically generates all the necessary code, so there is usually no need for extra work. However there are two points that the user should take care of:

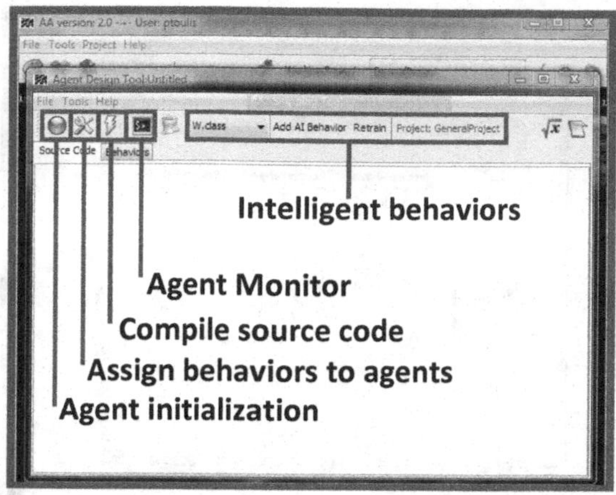

Fig. 5. Agent Design Tool features

1. Up to this point, there is no way for Agent Academy to infer the ontology the agents should communicate on. Therefore, the following source code line is added in the agent source code:

   ```
   this.getContentManager().registerOntology(**);
   ```

   The user, then, should replace the '**' with the desired ontology for the particular application. If a user-defined ontology does not exist, this line can simply be commented out.
2. Special attention should be given to the way AA handles behavior constructors. Currently, reflection is not used on behavior Java classes; as a result, only constructors with no arguments are assumed and at any other case users should manually edit the generated source code.

Given that all these issues are resolved, the generated source code can be compiled without problems.

A very important feature of ADT is the *Agent Monitor*, which can help greatly with the runtime debugging of the agent application under development. It is a graphical tool that enables AA developers to monitor when and in what order the behaviors of a specific agent are executed. Given that at the core of the agent-based software philosophy lies the multi-threaded (or parallel) programming model, this feature guarantees that agents will perform exactly as they were designed to. Additionally, the *Agent Monitor* provides with information on the message queue of agents, enabling the developer to inspect how messages are sent, received, and processed.

Figure 6 provides a screenshot of the *Agent Monitor* launched on an agent with three distinct behaviors. At any given time, only one behavior should be active, marked with a green square as shown in the figure. Using the monitor, the user may also inspect how behaviors are interchanged during execution, as well as the status and the message queue of the agent.

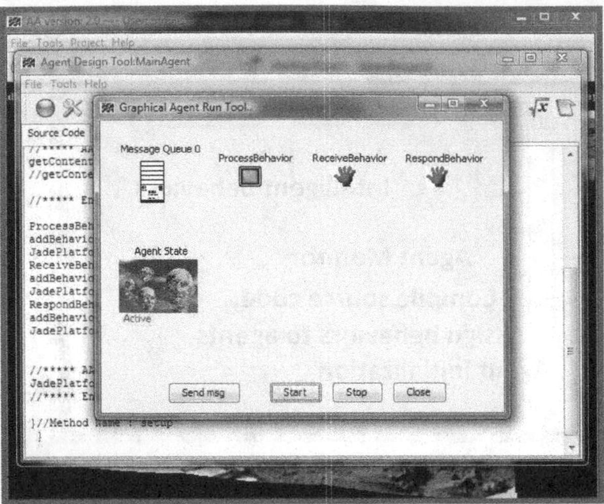

Fig. 6. The Agent Monitor

Another important feature of ADT is the intelligent DM behaviors toolbar. All available DM behaviors can be browsed in a drop down list and any of them can be added to the agent behavior pool with a simple click of a button. AA provides access to all intelligent DM behaviors that belong to the active project, so that it is possible for two or more agents to share them. When a DM behavior is retrained, it is dynamically updated to all agents employing it. And, though, agents of the same project may share behaviors that belong to that project, at the time being it is not possible for two projects to share code.

Multi-Agent System Design Tool. Having created all behaviors and all agents that implement the business logic of the application, the final step is to group everything under a Multi-Agent System. This is succeeded through the MAS design tool. AA saves all the information of the agents participating in the MAS in a simple text file with the extension *.aamas* (abbreviation for "Agent Academy MAS"), for latter use. In addition to defining the MAS, MDT initializes a JADE platform instance and allows the user to start/pause/stop the execution of the MAS. Parallel to the instantiation of the MAS, AA also loads the JADE sniffer, a tool for monitoring messages exchanged among agents.

4 Summary and Future Work

Within the context of this paper we have presented a methodology that provides the ability to *dynamically* embed DM-extracted knowledge to agents and multi-agent systems. Special emphasis is given on Agent Academy, the developed platform that supports the whole process of building DM-enhanced agent systems. It helps agent programmers to easily create agent behaviors, extract knowledge models based on Data Mining and integrate all these into fully working MAS.

AA can be thought of as a lower CASE (Computer Aided Software Engineering) tool, combining the power of two proven and robust software APIs, JADE and WEKA. Agent Academy is an open source IDE with features like structured code editing, graphical debugging and beginner-friendly interface to data mining procedures.

Future directions include the incorporation of the Agent Performance Evaluation (APE) [13] into the AA framework, in order to provide developers with the tools for evaluating the performance of the agent systems developed.

References

1. Adriaans, P., Zantinge, D.: Data Mining. Addison-Wesley, Reading (1996)
2. Bellifemine, F., Poggi, A., Rimassa, G.: Developing Multi-agent Systems with JADE. In: Castelfranchi, C., Lespérance, Y. (eds.) ATAL 2000. LNCS (LNAI), vol. 1986, pp. 89–101. Springer, Heidelberg (2001)
3. Cao, L.: Data mining and multi-agent integration. Springer-Verlag New York Inc. (2009)
4. Cao, L., Weiss, G., Yu, P.S.: A brief introduction to agent mining. In: Autonomous Agents and Multi-Agent Systems (2012)
5. Chen, M.-S., Han, J., Yu, P.S.: Data mining: an overview from a database perspective. IEEE Trans. on Knowledge and Data Engineering 8, 866–883 (1996)
6. Fayyad, U.M., Piatetsky-Shapiro, G., Smyth, P.: Knowledge discovery and data mining: Towards a unifying framework. In: Knowledge Discovery and Data Mining, pp. 82–88 (1996)
7. Foster, I., Jennings, N.R., Kesselman, C.: Brain meets brawn: why grid and agents need each other. In: Proceedings of the Third International Joint Conference on Autonomous Agents and Multiagent Systems, AAMAS 2004, pp. 8–15 (2004)
8. Gimpel, H., Jennings, N.R., Kersten, G., Okenfels, A., Weinhardt, C.: Negotiation, auctions and market engineering. Springer (2008)
9. Gruber, T.: Collective knowledge systems: Where the social web meets the semantic web. Web Semantics: Science, Services and Agents on the World Wide Web 6(1), 4–13 (2008); Semantic Web and Web 2.0
10. Han, J., Kamber, M.: Data Mining: Concepts and Techniques. Morgan Kaufmann Publishers (2001)
11. McCann, J.A., Huebscher, M.C.: Evaluation Issues in Autonomic Computing. In: Jin, H., Pan, Y., Xiao, N., Sun, J. (eds.) GCC 2004. LNCS, vol. 3252, pp. 597–608. Springer, Heidelberg (2004)
12. Kitchenham, B.A.: Evaluating software engineering methods and tool, part 2: selecting an appropriate evaluation method technical criteria. SIGSOFT Softw. Eng. Notes 21(2), 11–15 (1996)
13. Maimon, O., Rokach, L. (eds.): Soft Computing for Knowledge Discovery and Data Mining. Springer (2008)
14. Symeonidis, A.L., Mitkas, P.A.: Agent Intelligence Through Data Mining. Springer Science and Business Media (2005)
15. Witten, I.H., Frank, E.: Data Mining: Practical machine learning tools and techniques. Morgan Kaufmann, San Francisco (2005)

Role-Based Management and Matchmaking in Data-Mining Multi-Agent Systems

Ondřej Kazík[1] and Roman Neruda[2]

[1] Faculty of Mathematics and Physics, Charles University
Malostranské náměstí 25, Prague, Czech Republic
kazik.ondrej@gmail.com

[2] Institute of Computer Science, Academy of Sciences of the Czech Republic,
Pod Vodárenskou věží 2, Prague, Czech Republic
roman@cs.cas.cz

Abstract. We present an application of concepts of agent, role and group to the hybrid intelligence data-mining tasks. The computational MAS model is formalized in axioms of description logic. Two key functionalities — matchmaking and correctness verification in the MAS — are provided by the role model together with reasoning techniques which are embodied in specific ontology agent. Apart from a simple computational MAS scenario, other configurations such as pre-processing, meta-learning, or ensemble methods are dealt with.

Keywords: MAS, role-based models, data-mining, computational intelligence, description logic, matchmaking, closed-world assumption.

1 Introduction

An agent is a computer system situated in an environment that is capable of autonomous action in this environment in order to meet its design objectives [23]. Its important features are adaptivity to changes in the environment and collaboration with other agents. Interacting agents join in more complex societies, *multi-agent systems* (MAS).

The effort to reuse MAS patterns brings the need of separation of the interaction logic from the inner algorithmic logic of an agent. There are several approaches providing such separation and modeling a MAS from the organizational perspective, such as the *tuple-spaces*, *group computation*, *activity theory* or *roles* [6]. The Gaia methodology [25] fully exploits roles only in the analysis phase and leaves them during the design phase of development. The BRAIN framework [7]describes roles by means of XML-files and offers also the implementation support in JAVA language. The ALAADIN framework [10] is a organization-centered generic meta-model of multi-agent systems. It defines a general conceptual structure which is utilized in the MAS development. The framework describes MAS from an organizational perspective, instead of using terms of agents' mental states (agent-centered). This model (also called AGR) focuses on three basic concepts: agent, group and role.

L. Cao et al.: ADMI 2012, LNAI 7607, pp. 22–35, 2013.

Generally speaking, a role is an abstract representation of stereotypical behavior common to different classes of agents. Moreover, it serves as an interface, through which agents perceive their execution environment and affect this environment. Such a representation contains a set of actions, *capabilities*, which an associated agent may utilize to achieve its goals. On the other hand, the role defines constraints, which a requesting agent has to satisfy to obtain the role, as well as *responsibilities* for which the agent playing this role holds accountable. The role also serves as a mean of definition of *protocols*, common interactions between agents. An agent may handle more roles, and a role can be embodied by different classes of agents. Moreover, agents can change their roles dynamically. A group is a set of agents and its structure is defined by means of roles and protocols allowed in the group.

The role-based solutions may be independent of a particular situation in a system. This allows designing an overall organization of multi-agent systems, represented by roles and their interactions, separately from the algorithmic issues of agents, and to reuse the solutions from different application contexts. The coordination of agents is based on local conditions, namely the positions of an agent playing the role, thus even a large MAS can be built out of simple organizational structures in a modular way.

In order to automatize the composition of MAS, its formal model in description logic (DL) was introduced [16]. We are employing the concepts of role and group and transform the role model in axioms of DL [17]. In this paper, the necessity of axiom definition both under open- and closed-world assumption is highlighted and the model is extended by integrity constraints.

The main contribution of this paper is the unified formal model both for analysis and run-time support of the MAS which utilizes methods of automated reasoning. This formal description allows dynamic finding of suitable agents and groups (matchmaking), verification of correctness of MAS (system checking) or automated creation of MAS according to the task.

The *computational multi-agent systems*, i.e. application of agent technologies in the field of hybrid intelligence, showed to be promising by its configuration flexibility and capability of parallel computation, e.g. in [9]. We present a role-based model of complex data-mining scenarios, such as meta-learning, parallel computing, ensemble methods or pre-processing of data. The no-free-lunch theorem [24] shows that there is no best approach for every task. In practice the user does not know which method to use and how to set its parameters. Agent-based solution enables automatic assembly of the system. The previous experience can be stored for later experiments on similar data.

In the next section, we present a computational intelligence scenario, perform its analysis, and elaborate the role-based model of a computational MAS. In section 3, the model is formalized by means of description logic axioms. In section 4, the implementation of ontology agent managing the dynamic role-based of MAS, atomic actions and matchmaking queries, improving agents' sociability, are described. Section 5 concludes the paper and shows future work.

2 Role Model of Computational MAS Scenario

Hybrid models including combinations of artificial intelligence methods, such as neural networks, genetic algorithms, and fuzzy logic controllers, can be seen as complex systems with a large number of components and computational methods, and with potentially unpredictable interactions between these parts. These approaches have demonstrated better performance over individual methods in many real-world tasks [5]. The disadvantages are their bigger complexity and the need to manually set them up and tune various parameters.

There are various software packages that provide collection of individual computational methods, e.g. Matlab [12] or R Project [22] with focus on unified environment and computational efficiency. However, these frameworks are closed, they are inflexible in replacement of methods and they do not allow automated construction of data-mining processes. Moreover, it impossible to utilize their methods in various contexts such as meta-learning or recommending methods and settings based on previous experience. Other systems, such as the KNIME environment [4] focus on smooth user-friendly interface to complex hybrid models by offering visual assembly of data-mining processes.

Multi-agent systems seem to be a suitable solution to manage the complexity and dynamics of hybrid systems. In our approach, a computational MAS contains one or more computational agents, i.e. highly encapsulated objects embodying a particular computational intelligence method and collaborating with other autonomous agents to fulfill its goals. Several models of development of hybrid intelligent systems by means of MAS have been proposed, e.g. [15], [1] and [8].

In order to analyze computational scenarios and to construct a model of general computational MAS, we are exploiting the conceptual framework of the AGR model [10]. The simplest possible configuration is a task manager which controls model creation and data processing by the computational method — e.g. neural network — whose data are provided by the data source agent. The computational methods corresponds to physical implementation of agents employing the JADE agent platform and Weka data mining library [16]. This elementary computational scenario, where the computation consists of single method, is not sufficient for most cases we are interested in throughout this paper, but it is always contained in different contexts. Let us consider the following data-mining tasks:

- *Decentralized processing:* more computational agents controlled by single task manager, e.g. various ensemble methods or distributed execution of computational methods.
- *Supplementary learning:* separate optimization algorithms in the search space of computational methods' inner variables, e.g. optimization of neural network's weights by means of evolutionary algorithm.
- *Meta-learning scenario:* optimization in search space of method options [21].
- *Pre-processing and post-processing:* e.g. feature extraction, missing values and outlier filtering, or resampling etc. [11].

All these scenarios contain the coordination structures of computational agent which is controlled by a task manager. Also, the data are provided in all scenarios from a mediator, a pre-processing agent or final data source. In general, the machine learning of data-mining methods (e.g. a back-propagation trained perceptron network) can be seen as a search problem, either implicit — incorporated in the computational agent — or as an external search agent. The meta-learning scenario solves the optimization search of the method's options, thus the simple learning case is implicitly included there as an iterated subtask.

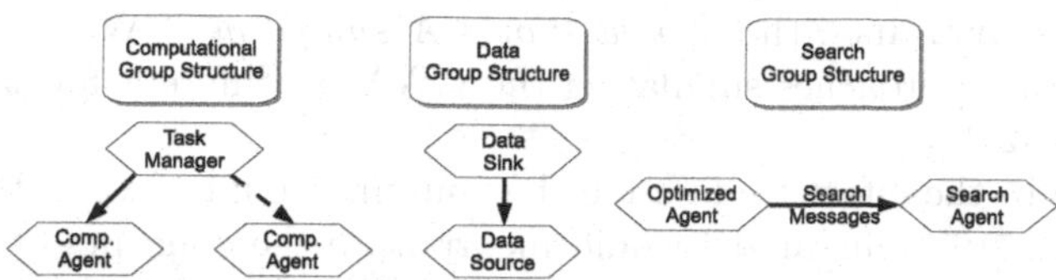

Fig. 1. The organizational structure diagram of the computational MAS

Three general subproblems are considered here: control of computational methods, data provision, and optimization search. Thus the decomposition results — according to the AGR model briefly described in section 1 — in a role *organizational structure* shown at Figure 1. It consists of possible groups, their structures, described by means of admissible roles and interactions between them. This organizational structure contains the following group structures:

- *Computational Group Structure.* It contains two roles: a Task Manager and Computational Agent implementing a computational method. The agent playing the Task Manager role can control more Computational Agents.
- *Search Group Structure* consisting of two roles: a Search Agent and Optimized Agent. It is representing search problem in a general search space.
- *Data Group Structure* contains a Data Sink and Data Source which provides it with data, e.g. training and testing data for computational agent.

The role model allows simplifying the construction of more complex computational multi-agent systems by its decomposition to the simple group structures and roles, which the agents are assigned to. Moreover, the position of an agent in a MAS in every moment of the run-time is defined by its roles without need to take account of its internal architecture or concrete methods it implements. It also reduces a space of possible responding agent when interactions are established, and the model will be used for matchmaking.

3 Description Logic Model of Computational MAS

The family of *Description Logic* (DL), fragment of first-order logic, is nowadays a de facto standard for ontology description language for formal reasoning [2]. In DL, a knowledge base is divided into a T-Box (terminological box), which contains expressions describing concept hierarchy, and an A-Box (assertional box) containing ground sentences.

Web Ontology Language (OWL), an expressive knowledge representation language, is based on description logic [20]. Semantics of OWL is designed for scenarios where the complete information cannot be assumed, thus it adopts the *Open World Assumption* (OWA). According to the OWA, a statement cannot be inferred to be false only on the basis of a failure to prove it. If the complete knowledge is assumed, the T-Box axioms cannot be used as *Integrity Constraints* (ICs) which would test validity of the knowledge base under OWA. In order to check integrity constraints, the *Closed World Assumption* (CWA) is necessary. There are several approaches simulating the CWA by different formalisms, e.g. rules or queries [20].

We continue in the effort to describe the computational MAS in the description logic model [16]. Our model would incorporate the concepts of group and role. In paper [17], we have elaborated basic role-based model of computational MAS in description logic under OWA. Limitation of standard OWL interpretation under OWA often leads to extension of description logic by other formalism (such as SWRL-rules [14]). We chose the solution where CWA axioms as integrity constraints are expressed in the same OWL language [20] since it preserves the model homogeneity. The authors presented an IC validation solution reducing the IC validation problem to SPARQL query [18] answering. Moreover, they introduced a prototype IC validator extending Pellet [19], the OWL reasoner. For example, the constraint that every service is provided an agent:

$$Service \sqsubseteq \exists isProvidedBy.Agent$$

would not be violated if there is defined a service without agent in an A-Box. The SPARQL representation of this IC would be the following query:

```
ASK WHERE {
  ?x rdf:type Service.
  OPTIONAL {
    ?x isProvidedBy ?y.
    ?y rdf:type Agent.
  }
  FILTER(!BOUND(?y))
}
```

Thus we divided the T-Box of the proposed model into two parts. The first part contains axioms describing mainly the concept hierarchy and the necessary relations between their instances. This schema is interpreted under the OWA and defines the facts the reasoner will infer from the given A-Box. In the second part, there are constraints which define the integrity conditions of the system related mainly to the capabilities of agents. These are interpreted under the CWA. Axioms of T-Box are distinguished in the following text by a subscript of the inclusion axiom symbol. A standard schema axiom interpreted under the OWA is in the form $C \sqsubseteq_O E_1$. An integrity constraint under the CWA has the form $C \sqsubseteq_C E_2$. The time-dependent information, the current state of the system, is in an A-Box of the ontology.

In the following, we describe a definition of the generic AGR concepts in DL. As we have already mentioned, a role is defined as a set of capabilities, i.e. actions (interactions) an agent assuming this role can use, and a set of responsibilities or events the agent should handle. A group is then described by a set of roles the group contains. A hierarchy of concepts should respect this. In the model, the running agents, groups and initiators are represented as individuals in A-Box. Their roles (i.e. sets of agents with common interface) and group-types are classes defined in T-Box statically.

The T-Box contains the following superior concepts (see Figure 2 left):

- *Responder* is a responsibility of a role. It stands for a type of interaction protocol the agent can handle.
- *Initiator* represents an action from a capability set, and it is coupled to a particular *Responder*. The functional role *isInitiatorOf* relates to the agent which the action uses. The role *sendsTo* contains the responding agents to which the action is connected.
- *RequestInit* is a subclass of the previous concept which defines only those initiators that send messages to one agent (unlike e.g. the contract net protocol).
- *Agent* is a class of all running agents and it is a superclass of all agents' roles. The role-agent assignment is achieved simply by a concept assertion of the agent individual in A-Box. The inverse functional roles *hasInitiator* (inverse of *isInitiatorOf*) and *hasResponder* couple an agent with particular actions and responsibilities. While the *hasResponder* relation is a fixed property, the *hasInitiator* occurs only when a corresponding connection is established. Finally, the functional role *isMemberOf* indicates belonging to a group.
- *Group* concept represents a group in a MAS. It has only an inverse of the *memberOf* role, called *hasAgent*. This relation has two subroles: functional *hasOwner* indicating the member agent which created the group and *rankAndFile* of all other member agents.

Now we begin to implement our domain specific groups and roles for our domain — computational MAS. The *computational group structure* contains agents with assigned roles of a task manager and one or more computational agent. Between the task manager and each computational agent a control interaction exists.

The sending of control messages between the task manager (*TaskManager*), which initiates this connection, and the computational agent (*CompAgent*) which performs the computational method is modeled by two concepts, an initiator (*ControlMsgInit*) and a responder (*ControlMsgResp*). The initiator of this connection is an instance of *ControlMsgInit* which is a subclass of the *Initiator* class. It sends messages only to an agent with a running responder handling these messages, and it is coupled with a Task Manager role as a capability. The schema file of the ontology contains axioms of the initiator concept hierarchy:

$$ControlMsgInit \sqsubseteq_O Initiator$$

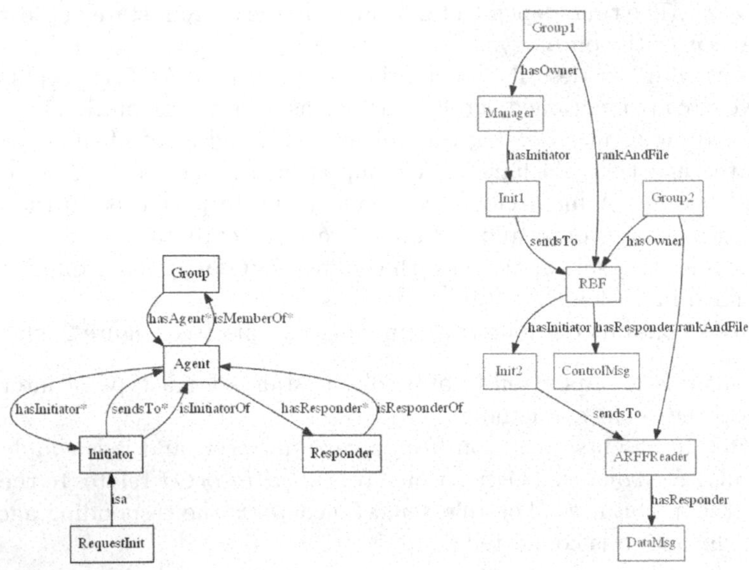

Fig. 2. *Left:* Superior concepts and their relations in T-Box of the role model. *Right:* A-Box state of the simple computational MAS configuration with computational and data group.

The control message responder is a simple descendant of the *Responder* concept and this class contains the instance *ControlMsg*. The schema axioms follow:

$$ControlMsgResp \sqsubseteq_O Responder$$
$$ControlMsgResp(ControlMsg)$$

The following integrity constraints for this concept check the roles of initiating and responding agents:

$$\begin{aligned} ControlMsgInit \sqsubseteq_C\ & \forall sendsTo.\exists hasResponder.ControlMsgResp \\ & \sqcap\ \forall isInitiatorOf.TaskManager \end{aligned}$$

Role definitions are descendants of the *Agent* concept and have to contain their responsibilities, i.e. responders (capabilities are defined on the initiator side). The responsibility of the computational agent (*CompAgent*) is to respond on the control connections. These are axioms inserted in the schema set:

$$CompAgent \sqsubseteq_O Agent \sqcap \ni hasResponder.ControlMsg$$

The task manager (*TaskManager*) role only sends messages in a group:

$$TaskManager \sqsubseteq_O Agent$$

Finally, the computational group (*CompGroup*) contains only the agents which have asserted that they have the computational agent, task manager or data source role. The subclass-axiom is important for open world reasoning:

$$CompGroup \sqsubseteq_O Group$$

On the other hand, admission (as an owner or general member) of the agent with a wrong role has to be checked by the following closed world constraint:

$$\begin{aligned} CompGroup \sqsubseteq_C \; & \forall hasAgent.(CompAgent \sqcup TaskManager) \\ & \sqcap \; \forall hasOwner.TaskManager \\ & \sqcap \; \forall rankAndFile.CompAgent \end{aligned}$$

The *data group structure* consists of two roles: the data source owning the group, and the data sink. Between the agents with these two roles is interaction providing data. The data sink requests for certain data the data source which sends them back. The following axioms for these concepts are similar to those for the computational group:

$$\begin{aligned} DataMsgInit \sqsubseteq_O \; & RequestInit \\ DataMsgInit \sqsubseteq_C \; & \forall sendsTo.\exists hasResponder.DataMsgResp \\ & \sqcap \; \forall isInitiatorOf.DataSink \\ DataMsgResp \sqsubseteq_O \; & Responder \\ & DataMsgResp(DataMsg) \\ DataSource \sqsubseteq_O \; & Agent \sqcap \ni hasResponder.DataMsg \\ DataSink \sqsubseteq_O \; & Agent \\ DataGroup \sqsubseteq_C \; & \forall hasAgent.(DataSource \sqcup DataSink) \\ & \sqcap \; \forall hasOwner.DataSource \\ & \sqcap \; \forall rankAndFile.DataSink \end{aligned}$$

The elementary computational scenario fits these two group structures. At Figure 2 right, there is A-Box state of the DL model. It contains two group individuals, *Group*1 which is instance of *CompGroup* and *Group*2 with type *DataGroup*. The depicted MAS consists of the following three agents: the Task Manager *Manager*, Data Source *ARFFReader*, and *RBF* agent implementing RBF Neural Network which has two roles: Computational Agent and Data Sink. The individuals *Init*1 and *Init*2 are two initiators realising the control and data connections.

The task manager is also able to control more computational agents in parallel and to collect their results. This configuration is shown at Figure 3 left in simplified cheeseboard diagram [10].

The *pre-processing agent*, i.e. encapsulation of a pre-processing method, obtains data from a data source and provides pre-processed data to other agents. The options of the pre-processing method and source-file have to be set by a

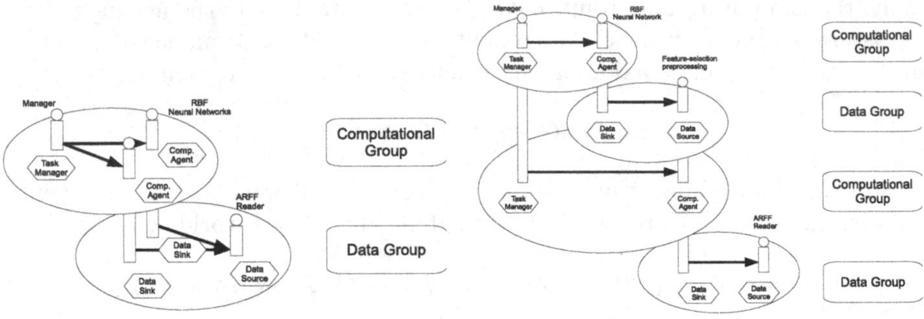

Fig. 3. *Left:* Example of computational MAS configuration with two computational agents controlled by single manager, which process the same data in parallel. *Right:* Example of computational MAS configuration with a pre-processing agent.

task manager who controls the computation. The pre-processing agent gains properties of both the data source (it provides data) and computational agent (it receives data from another source and waits for control messages). Thus the role of $PreprocessingAgent$ is defined as an intersection of $DataSource$ and $CompAgent$:

$$PreprocessingAgent \sqsubseteq_O CompAgent \sqcap DataSource$$

According to this definition, the pre-processing agent with this role is able to be controlled by a task manager in its own computational group and provide the processed data to another a data sink (e.g. computational agent) as a data source in data group. It also includes the possibility of creation of chain of agents, where on the one end is an agent providing original data table and on the other is a data mining computational method. Diagram of such a configuration with a pre-processing agent is at Figure 3 right.

The *search group structure* is defined in a similar way by the following schema and integrity rules:

$$\begin{aligned}
SearchMsgResp &\sqsubseteq_O Responder \\
& SearchMsgResp(SearchMsg) \\
SearchMsgInit &\sqsubseteq_O RequestInit \\
SearchMsgInit &\sqsubseteq_C \forall sendsTo.\exists hasResponder.SearchMsgResp \\
&\sqcap \ \forall isInitiatorOf.OptimizedAgent \\
OptimizedAgent &\sqsubseteq_O Agent \\
SearchAgent &\sqsubseteq_O Agent \sqcap \ni hasResponder.SearchMsg \\
SearchGroup &\sqsubseteq_O Group \\
SearchGroup &\sqsubseteq_C \forall hasAgent.(OptimizedAgent \sqcup SearchAgent) \\
&\sqcap \ \forall hasOwner.OptimizedAgent \\
&\sqcap \ \forall rankAndFile.SearchAgent
\end{aligned}$$

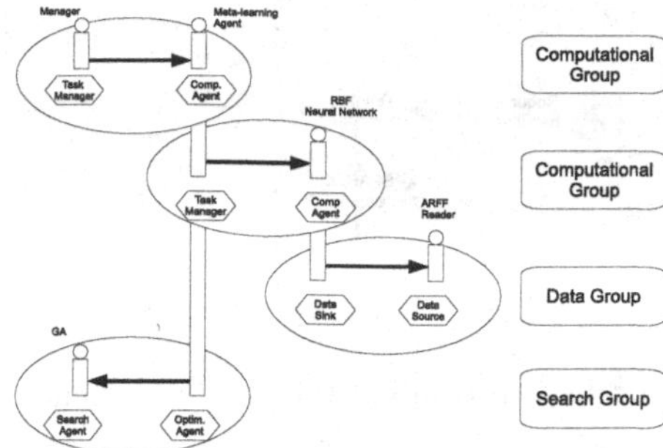

Fig. 4. The configuration of parameter-space search

This structure has various ways how it can be utilized. It can be used for optimization of computational agent's inner variables but also for meta-learning, i.e. finding of optimal options of certain computational method. In this case the agent does not control directly computational methods, but a meta-learning agent whose goal is finding the options with lowest resulting error rate. The meta-learning agent is optimized by a search method, and in the same time it controls the CAs representing the computational method. Such a configuration is shown at Figure 4. Here the options of the RBF neural network on certain data are optimized with Genetic Algorithm.

4 Management of the Model

To coordinate the run-time role organization of a MAS built according to the schemas and constraints of T-Box, it is necessary to have a central authority, separate agent in which the DL model is represented. Other agents will change the state of the model and query it by interaction with this agent.

The model is implemented as an *ontology agent* (OA) in JADE, Java-based framework for a MAS [3]. The goals of the OA are:

- Keeping track of the current state of MAS. Agents present in the MAS register themselves in the OA, state changes of their roles, create and destroy groups and their membership in them, and establish communication channels. The OA infers the necessary facts by means of OWA reasoning. The atomic actions are presented in subsection 4.1.
- Verification of correctness of MAS. The OA controls all changes of the system and does not allow activities which would violate the integrity constraints. The integrity constraints are handled by OA in CWA mode, as discussed in section 3.
- Matchmaking of agents and groups. When exploiting the concept hierarchy, it is possible to search groups of certain types or agents which have a certain role, which are members of certain group or that can handle certain types of messages etc. For the matchmaking queries, see 4.2.

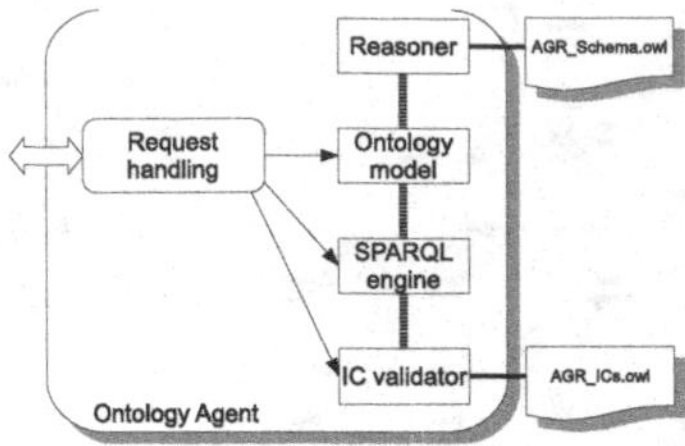

Fig. 5. Architecture of the ontology agent

The ontology agent (shown in Figure 5) consists of the *request handling* module which is responsible for processing of incoming requests and replying. It employs the ontology functions provided by the Pellet OWL-DL reasoner [19] and its extensions. The *ontology model* contains an assertional box of the ontology and describes the current state of the system. The open-world *reasoner* infers new facts from axioms in the OWL schema file and content of the A-Box. The integrity constraints saved in a separate OWL file are converted into SPARQL queries and run by the *SPARQL engine* on the ontology model. The SPARQL engine is also used to answer matchmaking queries.

The *communication ontology* for contents of OA messages has been created. This ontology consists of three types of concepts: *actions* changing state of the model, *matchmaking queries* and concepts informing about *results* of these requests.

4.1 Actions in the Model

During their life-time, the agents in the system will change their roles as well as position in the system. In order to affect the state of the role-based model, it uses several atomic actions which are sent to the ontology model. These actions result in changing of assertions in the A-Box of the model and are validated by the integrity constraints. If any of the ICs is violated, the change is not performed. In this case the action ends by failure.

The following types of actions are allowed in the MAS model:

- `RegisterAgent` — creates a new individual of type *Agent* and returns its name.
- `DeregisterAgent` — removes the individual which corresponds to the agent from the model.
- `SetRole(role)` — adds declaration of agent's role in the A-Box. E.g. if the agent a sets its role to computational agent, the OA adds in the A-Box the assertion $CompAgent(a)$.
- `RemoveRole(role)` — removes the role declaration from the A-Box.
- `CreateGroup(grpType)` — creates new individual defining group and declare it to type `grpType`. The relation *hasOwner* is created between the creating agent and the new group. Successful action returns name of the new group individual.

- `DestroyGroup(grpName)` — removes the individual `grpName` and all its assertions (including agents' membership).
- `EnterGroup(grpName)` — adds the relation *rankAndFile* between group `grpName` and the requesting agent. This change of state can be in contradiction with allowed roles in the group.
- `LeaveGroup(groupName)` — removes all relations *hasAgent*, *hasOwner* and *rankAndFile* between the agent and the group.
- `CreateInitiator(initType)` — creates an individual defining communication initiator. The concept assertion of this individual with class `initType` and role assertion *hasInitiator* between the agent and initiator individuals are added. The successful action returns name of the initiator. The action can contradict the assumptions about associations of roles and initiators' types.
- `DestroyInitiator(initName)` — removes the individual `initName` with all its assertions.
- `CreateConnections(initName, receivers)` — has as a result addition of sequence of *sendsTo* role assertions between initiator `initName` and the agents in the `receivers`. The action can be refused if the initiator type does not match with responders of any agent.
- `DestroyConnections(initName, receivers)` — removes all such assertions.

4.2 Matchmaking of Agents and Groups

The query engine of the model containing current MAS state is employed as a service for agents, which want to find suitable partners in the system for collaboration. The agents are also able to relate to the system by means of social concepts, e.g. groups or roles. The matchmaking requests of agents or groups are transformed by the OA into SPARQL queries [20] and executed on the inferred model.

The concept `getAgent(grpName, responderTypes, role)` specifies properties of demanded agent, i.e. the group it should be member of, list of communication types it should handle, and its role. For example, the matchmaking problem of finding of computing agent (i.e. has a type of *CompAgent*), which is member of group `g`, is converted to the following SPARQL query over the inferred A-Box of the model:

```
SELECT ?Agent WHERE {
  ?Agent rdf:type CompAgent.
  ?Agent isMemberOf g.
}
```

The ontology agent sends back in the inform message the list of agents matching with these properties.

Likewise, the request `getGroup(grpType)` returns list of groups in the MAS with specified type `grpType`.

5 Conclusions

In order to support the real-world data-mining processes, the models of computational group, data group, search group, and pre-processing agent have been included. The computational agent can stand for whole computational group and realize various ensemble methods. The meta-learning scenario as well as external learning can be described and implemented in this model by means of general search schema. The proposed model of pre-processing also allows defining chains of pre-processing agents gradually solving the input data inconsistencies.

The ontology agent representing the model of current MAS state has been implemented. The ontology agent allows general management, correctness verification and matchmaking of the MAS with concepts of agents, roles and groups. For this purpose, reasoning and querying of the DL model is employed.

Both the deduction axioms and integrity constraints are defined in the same formalism of OWL-DL with distinction of open-world and closed-world assumptions. So far, some of these ideas have been incorporated into an existing agent-oriented data-mining system [13] with promising results.

In comparison with other role based models, the proposed model supports analysis, design and implementation phase of development. Moreover it defines it in a formal model of the OWL-DL standard. This allows including of automated reasoning methods, and utilizing them in run-time management.

Further research will be put in ontology classification of computational methods, their parameters and input data. This model will broaden the possibilities of the model to express the computational MAS dynamics. This unified model will support the construction of computational MAS according to a demanded task, finding suitable methods and agents in the system, and possibly its visualization.

Acknowledgments. Ondřej Kazík has been supported by the Charles University Grant Agency project no. 629612 and by the SVV project no. 265314. Roman Neruda has been supported by the Czech Science Foundation project no. P202/11/1368.

References

1. Albashiri, K.A., Coenen, F.: Agent-Enriched Data Mining Using an Extendable Framework. In: Cao, L., Gorodetsky, V., Liu, J., Weiss, G., Yu, P.S. (eds.) ADMI 2009. LNCS, vol. 5680, pp. 53–68. Springer, Heidelberg (2009)
2. Baader, F., et al.: The description logic handbook: Theory, implementation, and applications. Cambridge University Press (2003)
3. Bellifemine, F., Caire, G., Greenwood, D.: Developing multi-agent systems with JADE. John Wiley and Sons (2007)
4. Berthold, M.R., et al.: KNIME: The konstanz information miner. In: Data Analysis, Machine Learning and Applications. Studies in Classification, Data Analysis, and Knowledge Organization, pp. 319–326. Springer (2008)
5. Bonissone, P.: Soft computing: the convergence of emerging reasoning technologies. Soft Computing - A Fusion of Foundations, Methodologies and Applications, pp. 6–18 (1997)

6. Cabri, G., Ferrari, L., Leonardi, L.: Agent role-based collaboration and coordination: a survey about existing approaches. In: Proc. of the Man and Cybernetics Conf. (2004)
7. Cabri, G., Ferrari, L., Leonardi, L.: Supporting the Development of Multi-agent Interactions Via Roles. In: Müller, J.P., Zambonelli, F. (eds.) AOSE 2005. LNCS, vol. 3950, pp. 154–166. Springer, Heidelberg (2006)
8. Cao, L.: Data Mining and Multi-agent Integration. Springer (2009)
9. Cao, L., Gorodetsky, V., Mitkas, P.A.: Agent mining: The synergy of agents and data mining. IEEE Intelligent Systems 24, 64–72 (2009)
10. Ferber, J., Gutknecht, O., Michel, F.: From Agents to Organizations: An Organizational View of Multi-agent Systems. In: Giorgini, P., Müller, J.P., Odell, J.J. (eds.) AOSE 2003. LNCS, vol. 2935, pp. 214–230. Springer, Heidelberg (2004)
11. Gibert, K., et al.: On the role of pre and post-processing in environmental data mining. In: International Congress on Environmental Modelling and Software – 4th Biennial Meeting, pp. 1937–1958 (2008)
12. Gilat, A.: MATLAB: An Introduction with Applications, 2nd edn. John Wiley and Sons (2004)
13. Kazík, O., Pešková, K., Pilát, M., Neruda, R.: Implementation of parameter space search for meta learning in a data-mining multi-agent system. In: ICMLA, vol. 2, pp. 366–369. IEEE Computer Society (2011)
14. Martin, D., Paolucci, M., McIlraith, S.A., Burstein, M., McDermott, D., McGuinness, D.L., Parsia, B., Payne, T.R., Sabou, M., Solanki, M., Srinivasan, N., Sycara, K.: Bringing Semantics to Web Services: The OWL-S Approach. In: Cardoso, J., Sheth, A.P. (eds.) SWSWPC 2004. LNCS, vol. 3387, pp. 26–42. Springer, Heidelberg (2005)
15. Neruda, R.: Emerging Hybrid Computational Models. In: Huang, D.-S., Li, K., Irwin, G.W. (eds.) ICIC 2006. LNCS (LNAI), vol. 4114, pp. 379–389. Springer, Heidelberg (2006)
16. Neruda, R., Beuster, G.: Toward dynamic generation of computational agents by means of logical descriptions. International Transactions on Systems Science and Applications, 139–144 (2008)
17. Neruda, R., Kazík, O.: Role-based design of computational intelligence multi-agent system. In: MEDES 2010, pp. 95–101 (2010)
18. Prud'hommeaux, E., Seaborne, A.: SPARQL query language for RDF. Tech. rep., W3C (2006)
19. Sirin, E., Parsia, B., Grau, B.C., Kalyanpur, A., Katz, Y.: Pellet: A practical OWL-DL reasoner. Web Semantics: Science, Services and Agents on the World Wide Web 5(2), 51–53 (2007)
20. Sirin, E., Tao, J.: Towards integrity constraints in OWL. In: OWLED. CEUR Workshop Proceedings, vol. 529 (2009)
21. Soares, C., Brazdil, P.B.: Zoomed Ranking: Selection of Classification Algorithms Based on Relevant Performance Information. In: Zighed, D.A., Komorowski, J., Żytkow, J.M. (eds.) PKDD 2000. LNCS (LNAI), vol. 1910, pp. 126–135. Springer, Heidelberg (2000)
22. Teetor, P.: R Cookbook. O'Reilly (2011)
23. Weiss, G. (ed.): Multiagent Systems. MIT Press (1999)
24. Wolpert, D.H., Macready, W.G.: No free lunch theorems for search. Tech. rep., Santa Fe Institute (1995)
25. Wooldridge, M., Jennings, N.R., Kinny, D.: The Gaia methodology for agent-oriented analysis and design. Journal of Autonomous Agents and Multi-Agent Systems 3(3), 285–312 (2000)

Incentivizing Cooperation in P2P File Sharing
Indirect Interaction as an Incentive to Seed

Arman Noroozian, Mathijs de Weerdt, and Cees Witteveen

Delft University of Technology, The Netherlands
{a.noroozian,m.m.deweerdt,c.witteveen}@tudelft.nl

Abstract. The fundamental problem with P2P networks is that quality of service depends on altruistic resource sharing by participating peers. Many peers freeride on the generosity of others. Current solutions like sharing ratio enforcement and reputation systems are complex, exploitable, inaccurate or unfair at times. The need to design scalable mechanisms that incentivize cooperation is evident. We focus on BitTorrent as the most popular P2P file sharing application and introduce an extension which we refer to as the indirect interaction mechanism (IIM). With IIM BitTorrent peers are able to barter pieces of different files (indirect interaction). We provide novel game theoretical models of BitTorrent and the IIM mechanism and demonstrate through analysis and simulations that IIM improves BitTorrent performance. We conclude that IIM is a practical solution to the fundamental problem of incentivizing cooperation in P2P networks.

Keywords: Incentives for Cooperation, Peer to peer coordination, BitTorrent.

1 Introduction

It is widely agreed upon that BitTorrent's *Tit-for-Tat* (TFT) mechanism is successful in incentivizing selfish autonomous peers to contribute to others as long as they are still in need of their peers' contributions. However, this is insufficient to incentivize peers to contribute after they have downloaded the files they seek. Consequently many mechanisms have been developed for P2P systems that incentivize cooperation beyond this natural limit of *direct* TFT as implemented in BitTorrent.

In this body of work we introduce an Indirect Interaction Mechanism (*IIM*) that extends the BitTorrent TFT. It takes advantage of mining a large data set for intersting and behavioral patterns. The observed pattern is that peers tend to download multiple files simultaneously (*multi-swarming*) or are in possession of files that have been downloaded in the past. This means that contrary to the standard BitTorrent protocol in which pieces of a single file are bartered for pieces of the *same* file, pieces of *different* files can be bartered for each other in a coordinated manner that better matches the supply and demand of files. Our data demonstrates this is a simple task. We observe that a large number of peers are able to use IIM with little coordination required.

While other approaches focus on reward/punishment or reputation mechanisms for better cooperation [24,26,9,17], IIM incentivizes cooperation without the need for such mechanisms (e.g. sharing ratio enforcement). We demonstrate that discriminative indirect interaction is not only possible in but that it incentivizes cooperation in game theoretic terms and also improves or does not degrade the performance of the standard

L. Cao et al.: ADMI 2012, LNAI 7607, pp. 36–50, 2013.

BitTorrent protocol. IIM is suitable for online bartering applications such as P2P enabled *Video on Demand* or Online Music services. Furthermore, economic experiments also hint at the strategy being effective [18,21].

Dealing with the complexities of distributed open systems such as BitTorrent and designing effective cooperation and coordination among peers is an area which can greatly benefit from techniques for mining the large data sets available.

In Section 2 we first give background information on the BitTorrent protocol. In Section 3 we then build a game theoretical model of BitTorrent to formally investigate the incentives problem. In Section 4 we explain the concept of indirect interaction and present our indirect interaction mechanism as a solution to the problem along with its game theoretical model. Next we give our experimentation and validation in Section 5: We give our data mining results in Section 5.1 demonstrating easy coordination of peer interactions, and we validate our hypotheses in Section 5.2. In Section 6 we discuss experimental economics studies that indicate our approach to be effective, and then explore some of the notable related work on incentives and other work on file sharing (in Section 7). Finally we conclude and present some ideas for future work in Section 8.

2 Background

BitTorrent is a file sharing protocol designed by Bram Cohen [2]. In BitTorrent peers *barter* bandwidth. Each file is divided into pieces and subpieces. Peers interact in rounds during which they can download (sub)pieces from their neighboring peers.

All content that can be downloaded has an associated *.torrent* file in which information about the content, size, SHA-1 hashes of file pieces, etc. are stored. The *.torrent* file also contains the address of a central server referred to as a *tracker*. The tracker is the entity responsible for introducing peers to each other, and keeping information about the peers. A peer will contact the tracker to request a list of peers (typically random list of size 50) which it can contact to download content. It will periodically contact the tracker for more peers. Peers that are downloading the same content from the same tracker are organized into a logical *swarm*.

BitTorrent recognizes two types of peers. Ones that are downloading a file: *leechers*, and ones that posses the entire file: *seeders*. Leechers follow the aforementioned procedure to download. Seeders wait to be contacted to upload content. They get nothing in return for uploading. On the other hand leechers are obliged by the protocol to reciprocate other leechers that provide content to them in a tit-for-tat fashion. As explained below, a leecher's reciprocation is not necessarily proportional to the amount of contributions received (e.g. it can reciprocate with less bandwidth than it actually received).

The TFT mechanism is implemented in BitTorrent's *unchoking* algorithm and has been the main driver behind Bit Torrent's performance and popularity. Each peer's available upload bandwidth is split into upload slots. The number of slots vary depending on the available upload bandwidth (typically equals 5). The unchoking algorithm assigns upload slots to preferred neighboring peers. At the beginning of each bartering round a peer assesses the amount of bandwidth that it has received from its neighbors and gives away its upload slots to the highest ranking contributors. Seeders assign uploads slots to the fastest downloading peers in a round robin fashion. In order to explore the bandwidth capacity of other neighbors, peers assign one of their upload slots to a random neighboring peer every few number of rounds. This is referred to as *optimistic unchoking*. Optimistic unchoking has a two fold purpose as it also allows newly arriving peers to receive a piece with which they can start the bartering process. This is referred to as *bootstrapping*.

Despite the success of the TFT mechanism [2,11], seeders are the neglected party in BitTorrent. They gain no utility by participating. As a result peers have no incentive to remain connected to a swarm after they have downloaded a file. A frequently observed behavior is that leechers leave a swarm as soon as they turn into seeders [15]. This phenomenon has lead to a drive towards various mechanisms for incentivizing cooperation such as reputation mechanisms and sharing ratio enforcement in private BitTorrent communities. Reputation systems suffer from problems such as whitewashing [5], communication overhead, sybil attacks [3], and have remained quite theoretical due to implementation complexity [15]. Private communities on the other hand suffer from exclusivity, imbalanced supply/demand and unfairness [7,15]. The need to design scalable incentive mechanisms that mitigate such problems is evident.

3 A Model of BitTorrent

In order to examine the incentives problem formally we have derived a novel game theoretical model of BitTorrent. Several key aspects of the protocol have directed us towards our choice of model the first of which is uncertainty. That is, a BitTorrent peer never knows whether it has uploaded with a high enough bandwidth in order to be reciprocated. This means that under standard game theoretical assumptions (rationality and utility maximization) BitTorrent peers would be maximizing their *expected* utility. A second aspect is interaction in rounds. At the end of each round peers will assess and readjust their strategies. Our model accordingly characterizes the utility of each player for every round. The third aspect is the protocol's design to prefer to interact with peers that have a higher upload bandwidth and rare pieces of the file. We capture these aspects of the BitTorrent protocol in our model which is based on a proposed framework by *Buragohain et al.* [1].

We characterize a peer's bandwidth contributions as a vector in which all neighbors are indexed and the values correspond to the bandwidth that was contributed to the peer with that index. The notation b_i denotes this characterization for a peer i. We have $b_i = (b_{i0}, \ldots, b_{i\|N_i\|})$ where N_i is the set of peer i's neighbors.

Peers can make certain contributions to each other which we characterize as the fraction of the file they possesses. Based on a peer i's pieces and another peer j's pieces the notation α_{ij} represents the fraction of the file that i can provide to peer j. If α_{ij} is relatively large there is a higher chance that peer i has a rare piece in which j is interested.

BitTorrent peers have a preference for rare pieces so we assume that each contribution has a certain value for a peer based on the rarity of the piece. We characterize this with a factor v_{ij} which denotes the value that peer i assigns to receiving a contribution from peer j, normalized such that a value of 1 represents the exact compensation for contributing the complete upload capacity.

However, the rarity of a piece is not the only factor influencing the choice of potential interaction partners. The amount of contribution of a peer in the previous round also

Table 1. Parameters of the Model

N_i	The set of peer i's neighbors.
b_i	The vector $(b_{i0}, \ldots, b_{i\|N_i\|})$, ($b_{ij}$ denotes the bandwidth peer i assigned to j).
v_{ij}	The value to peer i of a contribution from peer j to peer i.
α_{ij}	The fraction of the total pieces that peer i can provide to peer j.
$P(\alpha_{ij}, b_{ij})$	The probability of peer i receiving a contribution from j, having made a bandwidth contribution of b_{ij} while it has α_{ij} to provide to j.

determines the choice. We denote by $P(\alpha_{ij}, b_{ij})$ the probability of a peer i receiving a contribution from j to which it has made a bandwidth contribution of b_{ij} while it has an α_{ij} fraction of the pieces to provide. We refer to this probability as the *probability of reciprocation*. This probability $P()$ in fact can be seen to capture the history of interaction as a belief where all history is represented in the current state. We assume that the probability of reciprocation is a monotonically increasing function in both α_{ij} and b_{ij} and greater than zero (zero when a peer owns the entire file). The logic here is that contributing more bandwidth results in a higher chance of reciprocation and since the more pieces a peer can provide the more likely it is to have a rare piece, a higher α also results in a higher chance of reciprocation. Table 1 summarizes the parameters of our model.

A strategy of a peer i in our game theoretical model is equivalent to the vector b_i. Naturally the sum of peer i's bandwidth contributions to its neighbors cannot exceed its upload capacity. Using the notation b_{-i} to denote the strategy of peers other than i, we describe the expected utility of a peer i as follows:

$$u_i(b_i, b_{-i}) = -\|b_i\| + \sum_{j \in N_i} [P(\alpha_{ij}, b_{ij}) \times v_{ij} \times \|b_j\|] \quad (1)$$

Equation 1 states that the expected utility of i is equal to the sum over all neighbors j of the probability $P(\alpha_{ij}, b_{ij})$ that i receives a contribution from j multiplied by the value that i assigns to receiving the contribution (v_{ij}) and the bandwidth at which it receives the contribution, minus its own contribution $-\|b_i\|$. Note that we model i's expected share of j's bandwidth also within the probability $P()$.

According to this equation, a peer has to either increase its contribution, or the fraction of the pieces that it can provide to its neighbors in order to increase its utility. This is a well known fact about the BitTorrent protocol and how it operates. While the model is capable of explaining the basics of the protocol it is also capable of accounting for many observed phenomena of the protocol including, freeriding [5] (also see [13]), large view exploits [27], strategic piece revelation [12], minimum reciprocation winning uploads [22] and clustering of similar bandwidth peers [10]. For further details we refer the reader to [20].

Equation 1 clearly demonstrates the seeding incentives problem. The utility of a seeding peer is always negative because it does not have demand for file pieces. This problem needs to be addressed in order to be able to sustain the operation of P2P systems and avoid the tragedy of the commons.

4 Indirect Interaction

Seeding incentives in BitTorrent can be created by allowing *indirect interactions*, where a peer i supplies pieces to peer j who can supply pieces of another file to some other peer k, who in its turn can supply i. Such a setting requires peers to be involved in up/downloading multiple files, i.e., they are *multi-swarming*.

Definition 1 ***(Multi-Swarming Peer).*** *A peer p is a* ***multi-swarming peer*** *if it is online in swarms $S = \{s_1, s_2, \ldots\}$ where $|S| \geq 2$ and $\exists s_i \in S$ in which p is a leecher.*

The extension of BitTorrent presented in this section enables such interactions. We therefore refer to this protocol as the *Indirect Interaction Mechanism* (*IIM*).

4.1 Indirect Interaction Mechanism

IIM is a mechanism that relies on the tracker and its information to assist multi-swarmers in finding suitable parties to interact with in the same way that a tracker introduces new sets of peers in the standard BitTorrent protocol. Additionally, however, the tracker uses a supply/demand graph of active swarms.

Definition 2 ***(Supply and Demand Graph).*** *A Supply and Demand graph is a directed graph $G = (V, E)$ with edge labels, where V is the set of swarms, and the edges in E are defined as follows.*

1. ***(Seeder-Leecher)*** *There is an edge (s_1, s_2) in E if there exists a peer p that is a leecher in s_1 and a seeder in s_2. The edge is labeled with the set of all such peers p.*
2. ***(Leecher-Leecher)*** *There is an edge (s_1, s_2) as well as an edge (s_2, s_1) in E if there exists a peer p that is a leecher in both s_1 and s_2. Again the edges are labeled with the set of all such peers p.*

Any peer p that is a seeder in both $s1$ and $s2$ **(Seeder-Seeder)** does not occur in any of the labels on edges. Figure 1 gives an example of a supply/demand graph. This graph, denoted by G above, is stored as an adjacency matrix at the tracker. The IIM then works as follows.

1. Multi-swarming peers seeking indirect interaction announce all swarms they are participating in to the tracker and do a request.
2. The tracker updates the Supply and Demand graph, and then uses breadth-first search to find a limited number of cycles of small length.
3. The tracker introduces the peers of the found cycles to the requesting peer.
4. The requesting peer contacts the peer succeeding it in the cycle with the information it received from the tracker and requests indirect interaction.
5. All subsequent peers in the cycle will do the same until the request message arrives back at the originating peer.
6. Once a cycle has been established peers can unchoke connections with the same TFT mechanism that they use for normal BitTorrent connections with the additional consideration that unchoking the first next peer along the cycle depends upon the amount of bandwidth received from the previous peer in the cycle.

Run time operations of the tracker scale linearly with the number of requesting peers (since the length of the cycles is bounded by a constant), and quadratically in memory requirements.

4.2 The IIM Model

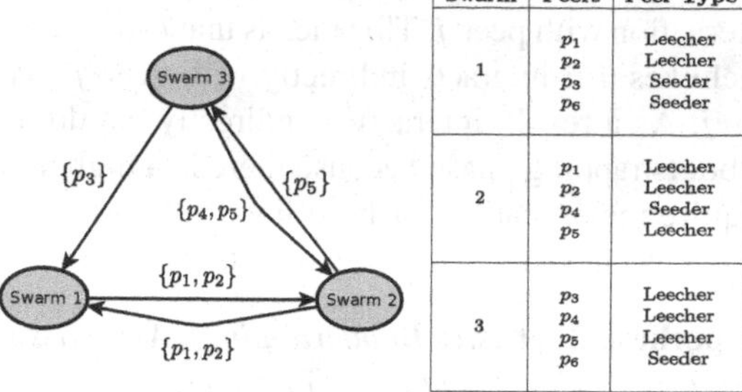

Swarm	Peers	Peer Type
1	p_1	Leecher
	p_2	Leecher
	p_3	Seeder
	p_6	Seeder
2	p_1	Leecher
	p_2	Leecher
	p_4	Seeder
	p_5	Leecher
3	p_3	Leecher
	p_4	Leecher
	p_5	Leecher
	p_6	Seeder

Fig. 1. Example of a Supply and Demand graph

For simplicity we only present the model of IIM for indirect interaction cycles of length 2. However, our model and analysis are generalizable to cycles of greater length.

Consider a peer i that is multi-swarming in two swarms. Let us assume that i has received a set of length 2 cycles C_i from the tracker and it decides to upload content to these cycles. We denote the contribution that i makes to each cycle by b_{ic} for $c \in C_i$. This creates a third term in the utility function as follows

$$u_i(b_i, b_{-i}) = -\|b_i\| + \sum_{j \in \hat{N}_i} [P(\alpha_{ij}, b_{ij}) \times v_{ij} \times \|b_j\|] + \sum_{c \in C_i} [P'(\alpha'_{ic}, b_{ic}) \times v_{ic} \times \|b_c\|] \quad (2)$$

Notice that peer i is making the same *total* contribution as before, i.e., $\|b_i\|$. From the perspective of the normal peers nothing has changed about the strategy of peer i because they derive expected utility based on i's total contribution. The interpretation of the second term in the equation is exactly the same as before. Notice that this time however, the probability of reciprocation from cycles (3rd term) depends on the fraction of pieces of the second file (α'_{ic}) that peer i can provide to individual peers in C_i. That is, peer i can provide α' pieces of one file in return for α pieces of the file it itself is interested in. Equation 2 constitutes our model for the proposed IIM mechanism.

Consider now a peer i which has a strategic choice between contributing through seeding and indirect interaction through a cycle of length two, or directly within the swarm it is leeching in (*i.e.* IIM *vs.* standard BitTorrent). For cycles of length two, we can safely assume that the probability P' of receiving a contribution through a cycle is the same as the probability of direct reciprocity P. However, when peer i is seeding, $\alpha_{ic} = 1$, so assigning the same bandwidth to the cycle is expected to contribute more to the utility than direct interaction. Consequently, a former seeder can now gain positive utility from seeding content instead of a negative utility as was the case in the original BitTorrent model (Equation 1).

4.3 IIM Properties

We now analyze the IIM model in a bit more detail. One consequence of the discussion above is that this also helps bootstrapping a peer. Consider a peer i that is partially or wholly in possession of one file and starting to acquire another file, while there are other peers j in possession of α and α' pieces of these files. At this stage of a download, $\alpha'_{ik} = 0 \; \forall k \in N_i$ (α is at its minimum) which means that the utility that peer i expects to derive from the second term of Equation 2 is at its lowest. Therefore i would have to wait to receive contributions as optimistic unchokes from its neighbors.

On the other hand, peer i possesses pieces of another file that it can use in indirect interaction with peer j. This means that i would no longer have to wait for the optimistic unchokes if it interacts indirectly with peer j. By symmetry the same argument holds for j. As a result, interacting indirectly would achieve a higher utility for both peers at bootstrapping phase because there is a higher chance that contributing in cycles will acquire reciprocation for both involved parties ($\alpha_{ij} > 0$ for peer i and $\alpha'_{ji} > 0$ for j). Hence we have the following hypothesis:

Hypothesis 1 ***(Faster Bootstrapping).*** *A multi-swarming peer will be able to bootstrap faster if it interacts indirectly in cycles.*

This time imagine a peer i which is leeching the file α and seeding the file α' (leecher-seeder) and another peer j which is doing the exact opposite (seeder-leecher). These two peers would gain additional utility by interacting indirectly in cycles because the fraction of the file that the pairs can provide to each other are maximal. Notice that if peer i already knows that peer j has chosen the option to interact indirectly with peer i, it would also decide to interact indirectly with peer j because it would derive a greater utility. The same argument is valid in reverse for peer j. As a result, peers i and j would at worst be indifferent to choosing either option.

Hypothesis 2 ***(Leecher-Seeder/Seeder-Leecher Pairs are Better Off Interacting Indirectly).*** *A multi-swarming peer i that is seeding file α' and leeching file α and a multi-swarming peer j that is doing the exact opposite will be better off to interact in cycles.*

Other combinations of the peers are less structured than the bootstrapping phase or the leecher-seeder/seeder-leecher pairs. These combinations are more difficult to derive analytically from the model. In such cases, the probability of reciprocation and the assigned value for each contribution from each peer have a more prominent role in determining the utility of this peer. However, by expectation interacting indirectly with a multi-swarming peer is similar to interacting with any other random peer in the set of neighbors. That is, we expect the fraction of pieces that the two multi-swarmers are able to barter and the values that they assign to each others' pieces be similar to those of a random peer in the set of their neighbors. Hence we can derive the following hypothesis regarding the other possible combinations of indirect interaction:

Hypothesis 3 ***(IIM Will not Reduce the Performance of the BitTorrent).*** *The standard BitTorrent protocol will not significantly outperform the indirect Interaction mechanism in terms of the time required to download an entire file.*

5 Experimentation and Validation

Our approach to validating claims consists of a measurement study and simulations of the indirect interaction mechanism. Our measurement study focuses on validating our assumption regarding the possibility of indirect interaction. It has helped us in designing IIM. Our simulations on the other hand focus on validating hypotheses that we derived from our game theoretic model of IIM. In short we demonstrate that IIM incentivizes potential multi-swarmers to choose indirect interaction over seeding in or disconnecting from swarms which could otherwise be used for indirect interaction.

5.1 Measurement Study

We have conducted a measurement study of a private BitTorrent tracker for the existence of potential supply/demand cycles. We have used the *FileList.org* community tracker data set previously studied by Roozenburg et al. [25]. Our choice of the data set has a two fold purpose. First, the possibility of indirect interaction is dependent on diversity and a large enough number of peers to locate *n*th parties. The FileList data is large enough to eliminate the possibility of not having enough diversity in supply and demand. Second, precise measurement of the possibility requires a data granularity at the level of individual peers. The FileList.org data contains information based on anonymized user names which allows us to uniquely identify users even if they reside behind a NAT device or Firewall or share the same IP address. The data set consists of a member base of some 110,000 peers forming some 2972 swarms. The key idea in our measurements is the notion of a *multi-swarming* peer.

The results of this measurement are depicted in Figure 2. Due to the complexity of finding all possible cycles of interaction we have limited our measurements to cycles of length no greater than 3 where a cycle of the form $p_a \rightarrow p_b \rightarrow p_a$ between peers p_a and p_b is considered of length 2 and a cycle of the form $p_a \rightarrow p_b \rightarrow p_c \rightarrow p_a$ is of length 3. Figure 2 reveals a surprising characteristic of the supply/demand structure within BitTorrent communities. As depicted there is a high probability that multi-swarming peers can interact in cycles of length 2 and 3. This means that for example there is a high probability that within the community a peer p_a that is leeching in swarm s_1 and seeding in swarm s_2 (leecher-seeder) can be matched with a second peer p_b that is doing the exact opposite by seeding in swarm s_1 and leeching in swarm s_2 (seeder-leecher).

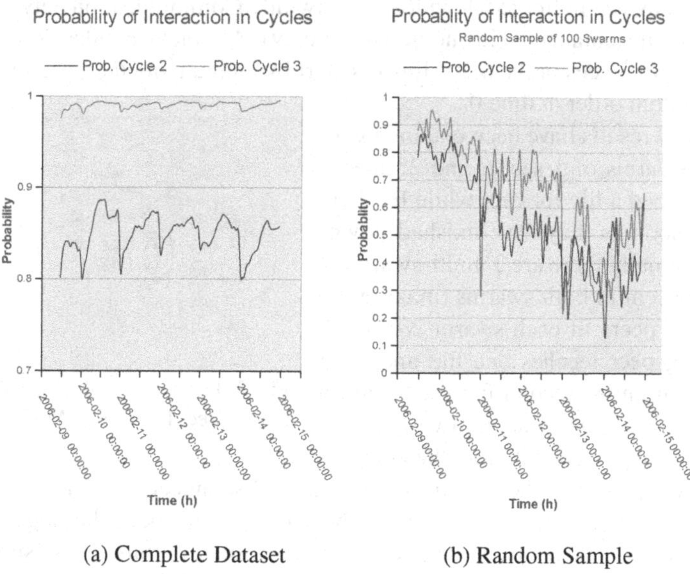

(a) Complete Dataset (b) Random Sample

Fig. 2. Probability of multi-swarming peers forming cycles of length 2 to 3

Given that the *FileList.org* community implemented weak sharing ratio enforcement it is possible that our measurements overestimate the potential for indirect interaction. In order to address this issue we have conducted a series of measurements by randomly choosing 100 swarms from the data set and repeating our measurements with the select few. The random nature and small sample size provide us with a fair estimation of how probable indirect interaction is without sharing ratio enforcement. A sample result is depicted in Figure 2. Similar results have been consistently observed. This demonstrates that sharing ratio enforcement does not affect our overall probability measurements. Potential cyclic interactions seem to be an intrinsic property of file sharing behavior. Other measurement studies [6] also suggest observation of peers downloading multiple files (up to 80%) at a certain moment which lead to multi-swarming. However such studies use IP addresses as peer identifiers which result in overestimation. Our data set contains much lower numbers of multi-swarmers (average of 30%).

Also note that to make an estimate of the effectiveness of IIM, we are not interested in how often peers currently are multi-swarming in public swarms, but in their potential to multi-swarm in case an incentive for seeding is given. The *FileList.org* community provides such an incentive by enforcing a (weak) sharing ratio, and is thus quite representative of expected results with IIM.

5.2 Simulations

In order to verify Hypotheses 1, 2 and 3 we have simulated IIM with a modified version of *TriblerSim* [17] Our modifications allow us to simulate a simple setting of IIM which involves two swarms and two multi-swarming peers (one seeder-leecher and one leecher-seeder) which can have a cyclic interaction. Figure 3 demonstrates the types of scenarios that our tool is capable of simulating.

The upload bandwidth of peers is assumed to be the bottleneck; therefore all peers are assumed to have an unlimited download bandwidth. Communication between peers is assumed to be instantaneous (i.e. no network delay). A typical simulation starts with leechers having no pieces of the file. All peers start by announcing their presence to the tracker in a random order at time 0.

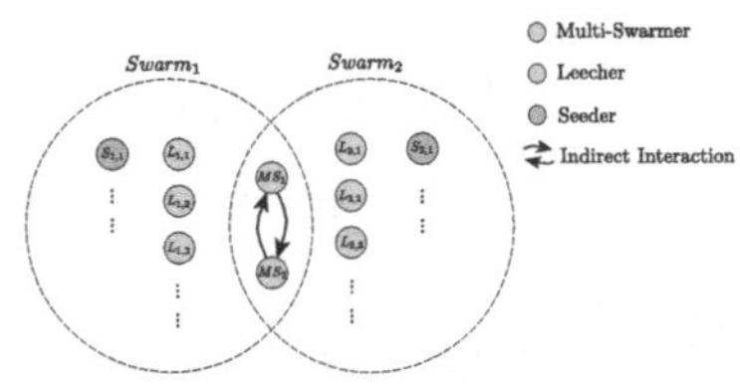

Fig. 3. Possible simulation scenarios of the Indirect Interaction Mechanism

Demonstrated results have been obtained with two identical swarms of 1 seeder and 25 leechers. Seeders upload a file for ever while leechers disconnect as soon as they have finished downloading. Additionally there are 2 multi-swarmers that that are present in both swarms (making the total number of peers in each swarm 28). Each multi-swarming peer leeches one file and is in possession of an entire second file such that an indirect interaction cycle of length two is possible. Multi-swarming peers withhold the contents of their second file from other leechers. They also disconnect as soon as they have finished downloading the file. All peers have the same upload speed (homogeneous environment). The results have been obtained with 50 simulation runs. These simulations are mirrored with the exact same number of runs in which the multi-swarmers do not interact indirectly (act as standard leechers).

Hypothesis 1 and 2. Figure 4 plots the bootstrap time of one multi-swarming peer using IIM against one using standard BitTorrent. We have measured the bootstrap time of a peer as the time required to download 10% of the pieces. Figure 4 demonstrates that a multi-swarming peer bootstraps faster by expectation (Hypothesis 1). Hence our hypothesis can be confirmed as our paired t-test comparison of means demonstrates: $\alpha = 0.01, t(df = 62.3769) = -11.1429, P = 0.0000$. We attribute this observed effect to the fact that a multi-swarming peers does not have to wait for optimistic unchokes to receive an initial piece and can directly barter a piece of a file that it possess for the first piece of the file that it seeks.

It turns out that our expectations regarding download time are also met. Similar to Figure 4 multi-swarmers manage to stay ahead of standard leechers. Hypothesis 2 is confirmed with a paired t-test comparison of means $\alpha = 0.01$, $t(df = 50.3766) = -32.5107, P = 0.0000$.

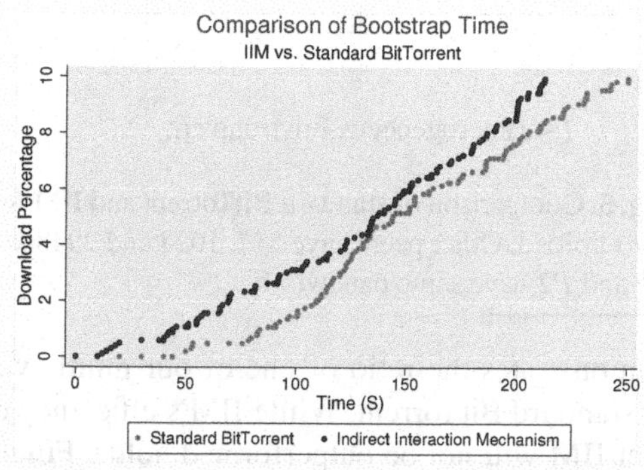

Fig. 4. Comparison of standard BitTorrent and IIM bootstrapping for the first multi-swarming peer

Hypothesis 3. In order to test our hypotheses more rigorously we have also simulated IIM in limited other settings. We only briefly report on some of our observations in such settings due to space limitations and refer the reader to [20] for more details.

We have conducted a series of simulations with three classes of peers with upload bandwidths of 2048, 1024 and 512*kbps*. In each simulation we have used two multi-swarmers as before, varied their class and compared their download times when using IIM and standard BitTorrent. Our observations indicate that when the classes of the multi-swarming peers are not too far apart or belong to the faster classes, the same advantages as before exist for using IIM even though one multi-swarmer is interacting with a slower multi-swarmer. This leads us to a conjecture that IIM improves download speed by expectation when the indirectly interacting parties have similar upload bandwidths (see Hypothesis 3).

An interesting class of results have been observed when the multi-swarming peers have the maximum difference in upload bandwidths. An example of such a simulation is demonstrated in Figure 5a. The observation here is that standard BitTorrent outperforms IIM in terms of download speed for the faster multi-swarmer but vice versa for the slower multi-swarmer. However, the performance loss for the faster peer is very small (rejected null hypothesis in t-test comparison of means). This indicates that even though download speed has been sacrificed, access to rare pieces has been able to compensate for much of the lost download speed. This indicates the possiblity of smart unchoking algorithms that can trade off between download speed and piece rarity in BitTorrent. Another series of our simulations in a homogeneous setting focuses on the effect of having more seeders in the swarm. The expectation here is that more seeders reduce the attractiveness of multi-swarming peers because of reduced piece rarity. Figure 6

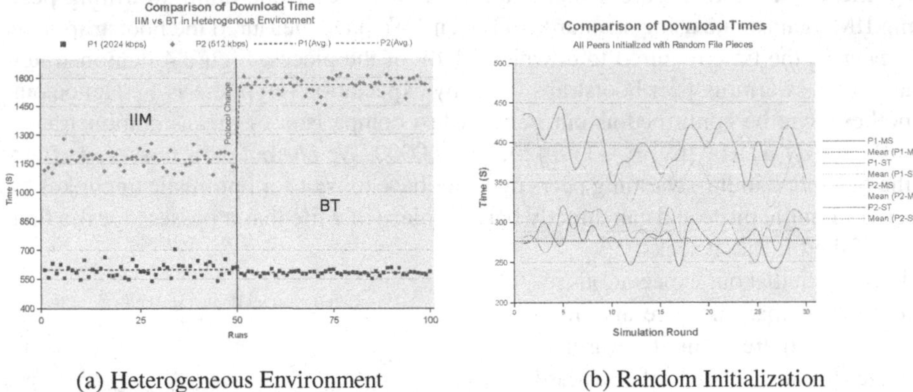

(a) Heterogeneous Environment (b) Random Initialization

Fig. 5. Comparison of standard BitTorrent and IIM for multi-swarming peers. 2 seeders with 2048 *kbps* upload; Other peers have 512, 1024 and 2048 *kbps* upload speeds. For Random Initialization $P1$ and $P2$ have same bandwidth.

demonstrates the ratio of one of our multi-swarming peer's download time with IIM to standard BitTorrent. While IIM's efficiency drops with more seeders, our hypothesis that IIM will not be outperformed holds. Finally Another series of simulation focuses on the scenario where the multi-swarming peers begin interaction at some later stage during their download process. This better matches real world scenarios. Figure 5b demonstrates the download time of two such peers. Starting interaction at a later stage gives the peers less time to capitalize on their access to rare pieces. As before our hypothesis holds. Here, P_1 and P_2 are the two multi-swarming peers. We compare their average download time with IIM and standard BitTorrent.

6 Discussion

BitTorrent (and generally P2P systems) have been studied game theoretically in numerous occasions [28,4].

One of the more popular models for P2P systems is the prisoners' dilemma (PD) game which captures a social dilemma between cooperation or defection among two players. In PD games the rational strategy is a non cooperative choice even though not the optimal choice. One of the differences between this and previous studies is that IIM allows us to study incentives from the perspective of a larger group than 2 players in the context of a more general game referred to as the Public Good game (PG).

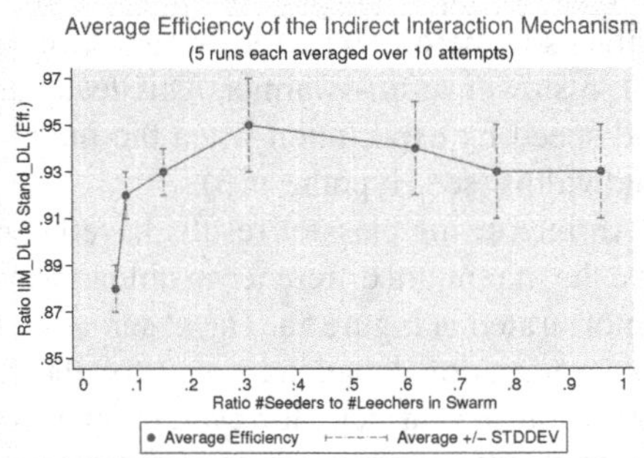

Fig. 6. Efficiency of IIM with increasing ratio of $\frac{seeders}{leechers}$. (MS = multi-swarming, ST= standard BitTorrent).

PG can be considered a more general form of PD. Its most common form consists of players which have been bestowed some amount of private good (tokens). The public good is represented by a public pot in which tokens are deposited. All players have to decide simultaneously on the amount of contribution that they would make to the pot. Once the contributions have been made, the pot is multiplied by some factor and divided equally among all players as a payoff for their investment.

Note the similarity between the utility in PG and the BT and IIM model. IIM can roughly be modeled as a repeated PG in which the initial tokens correspond to the peers's bandwidths. This allows us to derive a claim on user behavior from social experiments conducted with PG among a group of students by *Milinksi et al.* They show that people react positively towards *potential* reciprocation [18] (see also [21]). This experiment engages players in altering games of PG and indirect reciprocity. After some rounds of play groups are divided into two series in which one series knows that no more indirect reciprocity games follow. As a result the first group's contributions drop to zero while the second group player's continue to contribute. This suggests that as long as players have some expectation that indirect reciprocity will follow they still contribute for a series of rounds. Therefore we expect that with IIM a small probability of indirect interaction to be sufficient for people to seeding more often.

7 Related Work

Many incentive mechanisms appear in the literature among which, reputation systems, sharing ratio enforcement and private communities are the most notable.

Reputation mechanisms such as EigenTrust [9] are based on the concept of each peer having a global trust value that reflects the past experiences of other peers in the network with that peer. The main idea is to generalize local trust values to a global scale by weighting against the reputation scores of the assigning peers. Other approaches are based on only local trust values. There are practical issues with the storage and complexity of computing the reputation.

A second notable reputation mechanism is BarterCast [17]. BarterCast is a fully distributed sharing ratio enforcement protocol that uses an overlay network for peers to communicate their sharing ratio. An epidemic protocol is used for peer discovery and exchange of random lists of peers. A reputation metric and a Max-Flow algorithm are applied to the local view of each peer to gain an aggregate subjective view of the reputation scores.

Most reputation mechanisms are vulnerable to attacks ranging from simple lying about scores to more sophisticated attacks. Reputation systems generally cannot prevent such attacks [19]. Also they have remained quite theoretical and for example both EigenTrust and BarterCast face practical problems of communication overhead, collusion of malicious peers and sybil attacks [3]. A main advantage of IIM over reputation mechanisms is that IIM creates potential benefits almost in real-time, while reputation mechanisms only provide some promising future. We refer the reader to [14,18,4,5] for fundamental research on reputation systems.

Perhaps the most commonly used incentivizing mechanism is *sharing ratio enforcement* [17] in *private communities* [16]. This approach resorts to creating quantifiable metrics as a measure of cooperation. The cooperation metrics are used to create a standard of membership. Low standards result in disqualification. Despite all the advantages, private communities can lead to a phenomenon referred to as *BitCrunch* [7]. Informally, it refers to high bandwidth peers uploading most of the content while leaving little demand for the services of lower bandwidth peers. As a result it is becomes very hard for lower bandwidth peers to sustain high enough ratios to maintain memebrship or experience the benefits. This also leads to unfairness as these peers can experience long waiting times even though they are cooperative by definition and willing to share content. Another problem with private BitTorrent communities is their exclusivity.

Finally, other notable parts of the literature are analytical fluid models and game theoretical models of BitTorrent like systems. Fluid models [23,28] study the effect of churn and ratios of peer types on the properties of the systems and demonstrate that for certain scenarios Nash equilibria may exist that make uploading content more beneficial than freeriding. Other game theoretic approaches demonstrate strategies of better utilizing upload capacity [12,8].

8 Conclusions

We have outlined the Indirect Interaction Mechanism (IIM) as a mechanism for incentivizing seeding. The idea is that a peer can barter pieces of the file it has for pieces of the file it seeks. In addition, we show how this idea can even be extended to a cycle of contributions. IIM avoids most of the pitfalls of other incentivizing mechanisms: it is fair, it imposes reasonable overhead, and it can be used in public BitTorrent trackers. IIM creates close to real-time benefits for cooperating with other peers. The real-time nature of IIM makes it robust to malicious behavior.

The positive utility for seeders allows the use of the public goods game as a model for studying a group of peers. Social experiments with the game suggest that users react positively to indirect interaction given that some expectation of its possibility exist. Therefore, users are expected to choose IIM over standard Bittorrent.

Despite IIM being a centralized mechanism, it scales linearly with the number of multi-swarming peers [20]. The theoretically attractive solution would be to eliminate the centralized component in the system. There are two important aspects to decentralizing IIM: 1) Storage of Information regarding peer activities in swarms. 2) Finding potential indirect interaction partners.

A promising approach is the exploitation of gossip protocols such as ones utilized in BarterCast. Given a gossip protocol a multi-swarmer can individually discover potential partners in the gossiped information. We can speculate that there is a high chance that this type of information can be found because potential partners are both active within shared swarms and highly likely to receive information regarding each others' activities with low latency.

Distributed Hash Tables (DHTS) present another solution for the storage of information regarding peer activities in swarms. Peers would have to manually search the DHT for suitable indirect interaction partners.

References

1. Buragohain, C., Agrawal, D., Suri, S.: A game theoretic framework for incentives in P2P systems. In: IEEE Conf. on Peer-to-Peer Computing (2003)
2. Cohen, B.: Incentives build robustness in BitTorrent. Technical report (2003), bittorrent.org
3. Douceur, J.R.: The Sybil Attack. In: Druschel, P., Kaashoek, M.F., Rowstron, A. (eds.) IPTPS 2002. LNCS, vol. 2429, pp. 251–260. Springer, Heidelberg (2002)
4. Feldman, M., Lai, K., Stoica, I., Chuang, J.: Robust incentive techniques for peer-to-peer networks. In: Proc. of the 5th ACM Conf. on Electronic commerce, pp. 102–111 (2004)
5. Feldman, M., Papadimitriou, C., Chuang, J., Stoica, I.: Free-riding and whitewashing in peer-to-peer systems. In: Proc. of the ACM SIGCOMM Workshop on Practice and Theory of Incentives in Networked Systems, pp. 228–236 (2004)
6. Guo, L., Chen, S., Xiao, Z., Tan, E., Ding, X., Zhang, X.: Measurements, analysis, and modeling of bittorrent-like systems. In: Proc. of the 5th ACM SIGCOMM Conf. on Internet Measurement (2005)
7. Hales, D., Rahman, R., Zhang, B., Meulpolder, M., Pouwelse, J.: Bittorrent or bitcrunch: Evidence of a credit squeeze in bittorrent? In: 18th IEEE Int. Workshop on Enabling Technologies (2009)
8. Jun, S., Ahamad, M.: Incentives in bittorrent induce free riding. In: P2PECON 2005: Proc. of the 2005 ACM SIGCOMM Workshop on Economics of Peer-to-Peer Systems, pp. 116–121 (2005)
9. Kamvar, S.D., Schlosser, M.T., Garcia-Molina, H.: The eigentrust algorithm for reputation management in P2P networks. In: Proc. of the 12th Int. Conf. on World Wide Web, pp. 640–651 (2003)
10. Legout, A., Liogkas, N., Kohler, E., Zhang, L.: Clustering and sharing incentives in bittorrent systems. In: Proc. of the 2007 ACM Int. Conf. on Measurement and Modeling of Computer Systems, pp. 301–312 (2007)
11. Legout, A., Urvoy-Keller, G., Michiardi, P.: Rarest first and choke algorithms are enough. In: Proc. of the 6th ACM SIGCOMM Conf. on Internet measurement, pp. 203–216 (2006)
12. Levin, D., LaCurts, K., Spring, N., Bhattacharjee, B.: Bittorrent is an auction: analyzing and improving bittorrent's incentives. In: Proc. of the ACM SIGCOMM 2008 Conf. on Data Communication, pp. 243–254 (2008)
13. Locher, T., Moor, P., Schmid, S., Wattenhofer, R.: Free riding in bittorrent is cheap. In: SIGCOMM 2006 (2006)
14. Marti, S., Garcia-Molina, H.: Taxonomy of trust: categorizing P2P reputation systems. Computer Networks 50(4), 472–484 (2006)
15. Meulpolder, M.: Managing supply and demand of bandwidth in peer-to-peer communities. PhD thesis, Delft University of Technology (March 2011)
16. Meulpolder, M., D'Acunto, L., Capota, M., Wojciechowski, M., Pouwelse, J., Epema, D., Sips, H.: Public and private bittorrent communities: A measurement study. In: IPTPS 2010 (2010)
17. Meulpolder, M., Pouwelse, J., Epema, D., Sips, H.: Bartercast: Fully distributed sharing-ratio enforcement in bittorrent. Technical report, Delft University of Technology; Parallel and Distributed Systems Report Series (2008)
18. Milinski, M., Semmann, D., Krambeck, H.-J.: Reputation helps solve the 'tragedy of the commons'. Nature 415(6870), 424–426 (2002)
19. Nisan, N., Roughgarden, T., Tardos, E., Vazirani, V.V.: Manipulation-Resistant Reputation Systems. In: Algorithmic Game Theory, ch. 27. Cambridge University Press (2007)

20. Noroozian, A.: Incentivizing seeding in bittorrent - indirect interaction as an incentive to seed. Master's thesis, Delft University of Technology (2010)
21. Nowak, M.A., Sigmund, K.: Evolution of indirect reciprocity. Nature 437(7063), 1291–1298 (2005)
22. Piatek, M., Isdal, T., Anderson, T., Krishnamurthy, A., Venkataramani, A.: Do incentives build robustness in bittorrent. In: NSDI 2007 (2007)
23. Qiu, D., Srikant, R.: Modeling and performance analysis of bittorrent-like peer-to-peer networks. In: Proc. of the 2004 Conf. on Applications, Technologies, Architectures, and Protocols for Computer Communications, pp. 367–378. ACM, New York (2004)
24. Rockenbach, B., Milinski, M.: The efficient interaction of indirect reciprocity and costly punishment. Nature 444(7120), 718–723 (2006)
25. Roozenburg, J.: Secure decentralized swarm discovery in Tribler. Master's thesis, Delft University of Technology (2006)
26. Sigmund, K., Hauert, C., Nowak, M.A.: Reward and punishment. Proceedings of the National Academy of Sciences of the United States of America 98(19), 10757–10762 (2001)
27. Sirivianos, M., Park, J.H., Chen, R., Yang, X.: Free-riding in bittorrent networks with the large view exploit. In: 6th Int. Workshop on Peer-to-Peer Systems (2007)
28. Yu, J., Li, M., Wu, J.: Modeling analysis and improvement for free-riding on bittorrent-like file sharing systems. In: Int. Conf. on Parallel Processing Workshops (2007)

An Agent Collaboration-Based Data Hierarchical Caching Approach for HD Video Surveillance*

Wenjia Niu[1], Xinghua Yang[1], Gang Li[2], Endong Tong[1], Hui Tang[1], and Song Ci[1,3]

[1] Institute of Acoustics, Chinese Academy of Science
High Performance Network Laboratory
Beijing, China, 100190
{Niuwj,yangxh,tonged,tangh,sci}@hpnl.ac.cn
[2] School of Information Technology
Deakin University, 221 Burwood Highway
Vic 3125, Australia
gang.li@dekin.edu.au
[3] University of Nebraska-Lincoln
Omaha, NE 68182 USA

Abstract. In the research of networked HD video surveillance, the agent collaboration has been utilized as an emerging solution to collaborative caching in order to achieve effective adaption among the front-end HD video capture, the network data transmission and the data management for lossless video storage and complete playback. However, the cluster characteristic of various *caches* embedded in the *IP camera*, the *network proxy server* and the *data management server*, essentially contain important knowledge. How to utilize the cache clustering for collaborative stream controlling is still an open problem. In this paper, we propose an agent collaboration-based 3-level caching (*AC3Caching*) model, in which a cache *storage space*-based *AP* clustering mechanism is developed for fast grouping of "similar" caches on different levels. Furthermore, based on the cache cluster, transmission planning is designed based on the agent collaboration and reasoning. The experimental evaluations demonstrate the capability of the proposed approach.

Keywords: Video Surveillance, Agent Reasoning, Planning, Cache Cluster, Stream Control.

1 Introduction

Among the emerging research of networked HD video surveillance, it is not trivial to achieve effective adaption among the front-end HD video capturing, the

* This work was partially supported by the *National Natural Science Foundation of China* (No. 61103158,60775035), the *Strategic Pilot Project of Chinese Academy of Sciences* (No. $XDA06010302$), the *National S&T Major Special Project* (No. $2010ZX03004$-002-01), and the *Securing CyberSpaces Research Cluster* of Deakin University.

L. Cao et al.: ADMI 2012, LNAI 7607, pp. 51–64, 2013.

network data transmission and the data management for lossless video storage and complete playback. On one hand, HD camera will continuously generate large amount of data stream. If the network does not satisfy some pre-designed transmission bandwidth as a thick "pipeline", the overflow of video data will render the surveillance system inoperative. On the other hand, as we know, the intrinsic dynamic characteristics of the network (e.g. the *congestion* and the *interruption*) also require the transmission to be dynamically adjustable to allow correct data *storage* [9]. To address the adaption problem, the *caching* idea was gradually exploited in many efforts [15] [18] [16]. In other words, various *cache*s embedded in the *IP camera*, the *network proxy server* and the *data management server*, should be built and collaboratively used for controlling the data stream in HD video surveillance.

To help design collaborative caching mechanism, the *multi-agent* technology, which has provided methods of building agents capable of making autonomous decisions and also enabling the cooperation between those agents with mining ability, has been successfully utilized for video transmission [17] [19]. This is mainly for two reasons. Firstly, agents can proactively sense the latest dynamic changes on resources such as storage space and transmission bandwidth through their inherent autonomous computing. Secondly, through strong reasoning ability with *data mining* techniques such as *clustering* [20], agents can proactively cooperate with each other, and this provides the potentials for improving the accuracy and intelligence.

However, in real-world HD video surveillance applications, specific events (e.g. crime of affray) usually generate significant background changes, which may result in large output stream. While on most occasions, these events occurs in specific area, and it means that the *cache*s of *IP cameras* within this area could have similar value on their storage space (e.g. 10% to 15% *remaining space*) to form a *cluster characteristic* [5]. Similarly, the *network proxy server* and the *data management server* could also have the *cluster characteristic*. Hence, the cluster characteristic of various *cache*s embedded in *IP camera*, the *network proxy server* and the *data management server*, essentially contain important knowledge.

As we know, in general video surveillance process, there involve three kinds of knowledge flows. The *info flow* contains the collected necessary device or environment information, the *control flow* contains the camera control command, while the *video flow* contains the video data stream. For HD video transmission, transmission latency in the *video flow* could lead to severe transmission congestion and incomplete playback. If in advance we can utilize the above mentioned *cluster characteristic* of caches from the *info flow*, the stream transmission will be guided promptly so that the corresponding latency can be reduced.

In this paper, we will propose an agent collaboration-based data hierarchical caching approach for HD video surveillance. By utilizing the *cluster characteristic*, we develop 3-level hierarchical stream control for video transmission through agent collaboration-based caching. Focusing on the *video flow*, the *storage space* will be mainly utilized to distinguish different-level cache cluster characteristics. The main contribution of this work can be summarized as

- A cache *storage space*-based clustering mechanism for fast grouping of different-level "similar" caches into cluster, together with the dynamic cluster scheduling mechanism.
- Based on the cache cluster, we utilize agent collaboration to build a 3-level cache management model, in which agent reasoning-based transmission planning algorithm is proposed as well.

The rest of the paper is organized as follows. Section 2 discusses related work, followed by the proposed approach as described in Section 3. Section 4 presents experiments that illustrate the benefits of the proposed scheme. Section 5 concludes the paper.

2 Related Work

As we know, video transmission is the key issue in video surveillance. To increase the flexibility and intelligence in video transmission, the agent technology has been incorporated into many applications. Existing methods either extend the single agent's ability to accommodate specific functions (e.g. *caching*) or use multi-agents to build effective collaboration mechanisms. Along the first line of research, in the work [4], feedback adaptation has been the basis for video streaming schemes whereby the video being sent should be adapted according to feedback on the channel. Through the agent *IO* interface extension, they designed a streaming agent (*SA*) at the junction of the wired and wireless network, which can be used to provide useful information in a timely manner for video adaptation. The work [17] proposed a cost-effective solution called *caching agent* to enable an effective *video-on-demand* services on the Internet, in which agent can be equipped with a chunk-based local storage to facilitate caching. Along the second line of research, the work [10] presented a *MobEyes* middleware solution to support proactive urban monitoring applications, where the agents are used to harvest metadata from regular vehicles. Especially, their multi-agent coordination mechanism is developed with a priori knowledge of the location of the critical information based on biological inspirations such as foraging and so on. In our work at the agent aspect, in addition to agent collaboration design, how to extend agent with special reasoning ability for different-level caching control is a challenging task.

For common caching control, different kinds of caches have been exploited in video transmission applications. Existing methods mainly focus on the transmission optimization or the accuracy of content caching for inter-ISP bandwidth savings. The work [15] proposed a proxy caching mechanism, which utilized a prefetching mechanism to support higher quality cached streams during subsequent playbacks for layered-encoded video streams to maximize the delivered quality of popular streams to interested clients. The work [14] proposed a so-called multicast cache(*Mcache*). By using regional cache servers deployed over many strategic locations, *Mcache* can remove the initial playout delays of clients in multicast-based video streaming. The work [18] addressed the problem of efficient streaming a set of heterogeneous videos from a remote server through a proxy to multiple asynchronous clients so that they can experience playback with low startup delays, in which an optimal proxy prefix cache allocation to the videos that minimize the aggregate network bandwidth cost is developed. The work [16] designed the transcoding-enabled caching (*TeC*) to enable network proxies with the transcoding capability, and provide different, appropriate video quality to heterogeneous networks. The work [13] presented a hot-point and multivariate sector caching to store more videos without replacement policies, which placed hot-points at major and inner sub-levels in the video. Compared with above related work, our work will follow the research line of video transmission optimization rather than the content caching accuracy study. Especially, when the clustering technique is introduced into hierarchical caching control, the multi-granularity caching usage becomes very challenging and should be focused.

Moreover, *data mining* can be also utilized by agents to facilitate agent-based services [8] [1] [2] [11] [3]. Krishnaswamy et al. [8] present a hybrid architectural model for *Distributed Data Mining* (DDM), which is customized for e-businesses applications such as service providers selling *DDM* services to *e-commerce* users and systems. More recently, Cao et al. [1] [2] [3] contribute a series of great efforts on agent-mining based services. In their work, the interaction between agent technology and data mining presents prominent benefits to solve challenging issues in different areas. These researches on agent and data mining interaction (*ADMI*) have greatly inspired us and provided a solid support for building affinity propagation (*AP*) clustering [6] into agent architecture for HD video surveillance.

3 *AC3Caching*: An Agent Collaboration-Based 3-Level Caching Model

3.1 *AP* Clustering-Based Cache Organization

Recently, the *affinity propagation* (*AP*) algorithm [6] proposed by Frey and Dueck has been widely used and its effectiveness has been also proved in different applications. Here, we exploit the *AP* algorithm to dynamically group caches in video surveillance system. Different with many other clustering algorithms (e.g. K-means), the *AP* algorithm can simultaneously take all data points as potential *exemplars*. The similarity $s(i,k)$ indicates how well the data point k is suited to be the *exemplar* for data point i. For simplicity, we will regard each cache as a data point, which can be named Nc in short. There exist two kinds of real-valued messages for transmission in the AP algorithm: the *responsibility* and the *availability*.

In the *AP* clustering process, firstly, the *availabilities* are all initialized to *zero*, then the *responsibility* and the *availability* messages are exchanged between Nc_i and Nc_j. The message-passing procedure may be terminated after a fixed number of iterations, after changes in the messages fall below a threshold, or after the local decisions stay constant for some number of iterations.

On each layer, the clusters are made by the *AP* algorithm and maintained as local clustering information (*LCI*) at a pre-specified server of this layer.

Furthermore, a central *data management server* is needed to be set to collect *LCI*s of all the three layers and manage them by using a unified scheme, which will bring a hierarchical global cluster view of cache storage usage. This scheme contains the following parts: *layer name* specifies which layer the cluster information is from, i.e. *LCI* ID; *cluster number* specifies the local cluster size by an integer number K; *cluster ID* specifies the cluster index in the local layer by a unique ID i; *cluster node set* contains all the nodes of a cluster. Here, we utilize the data structure *ArrayList* S_i to store nodes in an array-based memory space, and each node will be further represented as a data structure *Vector*. The *node struct* specifies the elements of nodes in a vector data structure. In this vector, $Nc_q = (IP_q, Port_q, Ratio_q)$, where IP_q,$Port_q$ are used for effective *IP* access, and $Ratio_q$ reflects the remaining ratio of total storage space in Nc_q. The *cluster exemplar* specifies the exemplar Nc_e of S_i.

Due to the remaining ratio of cache storage will vary along with the video streaming transmission, the clustering results could be different with the time variation. As for the *LCI* updating issue, we can set a trigger with frequency $f = 1/\Delta t$ to activate the new-round clustering, and the experiment results in Sec. 4 will be provided together with further discussion.

3.2 Agent Collaboration Mechanism

Agent Roles. From the process perspective of video surveillance, three roles including video (*Capturer*, *Transmitter* and *Manager*) are involved. Accordingly, three types of agents will be needed: the *Capturer Agent* (CA), the *Transmitter Agent* (TA) and the *Manager Agent* (MA). Among them, the *Capturer Agent* provides front-end video capturing by *IPC*, the *Transmitter Agent* will be responsible for transmitting video stream from the *Capturer Agent* to the *Manager Agent*, and the *Manager Agent* will receive, store the video data and provide surveillance functionalities such as *video playback*.

While from the flow type perspective of video surveillance, for each *CA*, *TA* and *MA*, the *info* agent, the *control* agent and the *video* agent should be further classified in refined level respectively. We can use the symbol $\mathcal{P}_\mathcal{Q}, \mathcal{P} \in \{CA, TA, MA\}, \mathcal{Q} \in \{info, control, video\}$ to distinguish them. Among them, the *info Agent* will be responsible for the *info flow*-related processing. The *control Agent* will be responsible for the *control flow*-related processing, and the *video Agent* will be responsible for the *video flow*-related processing.

The collaborative relationship among agents with different roles is shown in Fig. 1. Corresponding physical location of simulated network is shown in Fig. 2, which will be explained in detail in Sec. 4. In addition to the interactions among *CA*, *TA* and *MA* as colored arrow shown, each interaction among $\mathcal{P}_{info}$,$\mathcal{P}_{video}$ and $\mathcal{P}_{control}$ (see labeled arrow) will be explained in detail as follows. In $L1$, CA_{info} can obtain *IPC*-related information (e.g. cache usage). In $L2$, $CA_{control}$ puts control command on *IPC*. In $L3$, CA_{video} not only obtains video data from *IPC*, but also do the video data fusion for transmission recovery. In $L4$, on one hand, CA_{info} needs to promptly achieve video streaming information from

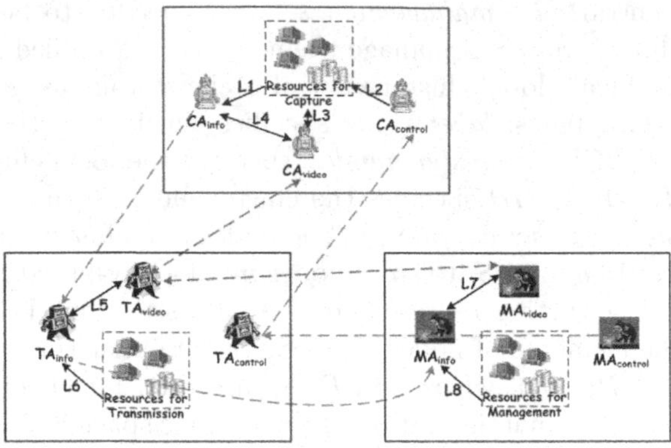

Fig. 1. Logical Agent Interaction

CA_{video}. Moreover, CA_{info} should provide its information to CA_{video} for cache control. $L5$ and $L7$ are very similar to the interaction between CA_{info} and CA_{video}. In $L6$, TA_{info} can obtain network-related information (e.g. network congestion and so on). In $L8$, MA_{info} can obtain *data management server*-related information (e.g. remaining storage space). Furthermore, we can see that, CA_{info}, TA_{info} and MA_{info} form a positive one-way *info flow*, $CA_{control}$, $TA_{control}$ and $MA_{control}$ form a reversed one-way *control flow*, while CA_{video}, TA_{video} and MA_{video} form a complete two-way *video flow*.

Agent Design. The detailed description about each agent's internal function modules is given as follows.

a) $\mathcal{P}_{info}$ Agent. The $\mathcal{P}_{info}, \mathcal{P} \in \{CA, TA, MA\}$ agent mainly provides the basic information acquisition, fusion and maintenance. To be noted that, compared with other agent roles, the cache clustering will be executed at $\mathcal{P}_{info}$ and corresponding *LCI* is maintained as well. As shown in Fig. 3(a), from the top to bottom, we can see that there exist the *interface* module, the *fusion* module, the *crawler* module, the *clustering* module, the *reasoning* module, the *basic storage* module, the *LCI storage* module and the *rule lib* module.

The *interface* module is responsible for agent communications inside and outside by passing standard agent communication language (*ACL*). More specially, it can receive any request from other agents, then the agent will execute corresponding functions and finally return the results still through the *interface.* The *fusion* module is responsible for making necessary information fusion among CA_{info}, TA_{info} and MA_{info}. We can regard it as a *porter*, who will build a bridge for connecting different-level agents and make the *info* flow get flowing. For example, it can collect all the *LCI*s to form a global cluster knowledge. The *crawler* module is responsible for obtaining the *info* data from other agents or environments, and then pushing these knowledge into the *basic storage* module. The *clustering* module can drive the *crawler* to capture updated *info* data. Then according to the knowledge in the *basic storage* module, it can execute *AP* clustering algorithm. The clustering results *LCI* will be stored into the *LCI*

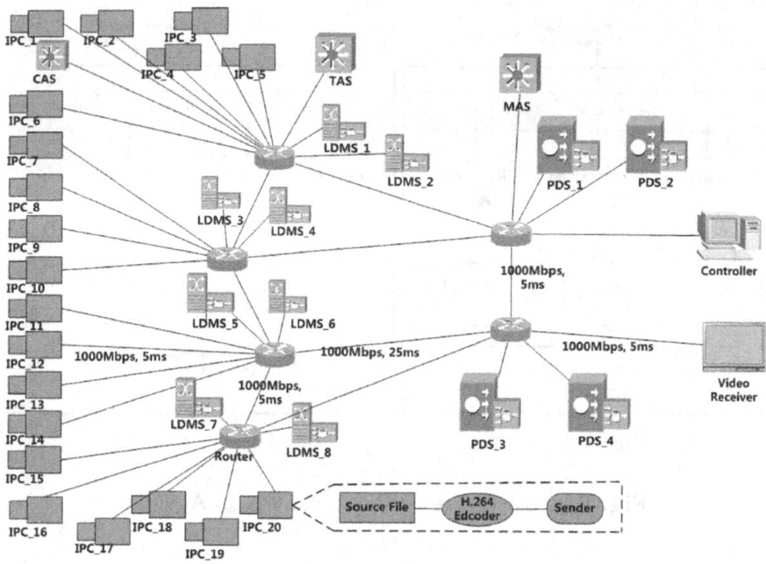

Fig. 2. Physical Agent Location in Simulated Network

storage module. Based on the rules in the *rule lib* and the *LCI*, the reasoning module can be used to decide which clusters should be utilized at current stream transmission path. Moreover, for those clusters on the stream transmission path, we need further decide which caches should be utilized.

b) $\mathcal{P}_{video}$ **Agent.** As shown in Fig. 3(b), the $\mathcal{P}_{video}$ agent has similar architecture with the $\mathcal{P}_{info}$ agent. Their differences can be summarized as follows: firstly, the $\mathcal{P}_{video}$ agent does not have the *clustering* module or the *LCI storage* module, which means that the $\mathcal{P}_{video}$ agent will not be responsible for clustering. Secondly, as the agent for the *video flow*, cache will be deployed in the $\mathcal{P}_{video}$ for video stream transmission. Thirdly, different with the *reasoning* in the $\mathcal{P}_{info}$ agent, the *planning* in the $\mathcal{P}_{video}$ agent is responsible for giving a specified effective transmission path and guiding the *video flow*. Please be noted that, the *planning* closely need the support of the results of *reasoning*.

c) $\mathcal{P}_{control}$ **Agent.** As shown in Fig. 1, the $\mathcal{P}_{control}$ agent has the most simple interaction with other agents. Hence, it can be designed to only have two modules, i.e., the *basic storage* module and *the fusion* module. With these two modules, the $\mathcal{P}_{control}$ agent can build a bridge for connecting different-level agents and enable the end-to-end arrival of camera control command.

Cache Cluster-Oriented Agent Reasoning. For these cache clusters, two kinds of reasoning should be developed: one is to decide which clusters should be utilized on the current stream transmission path; the other is to decide which caches of the cluster should be utilized. Considering the efficiency requirements of fast reasoning, we mainly exploit the rule matching-based reasoning rather than logic-based reasoning, and adopt the *Rete* [7], an efficient forward inference

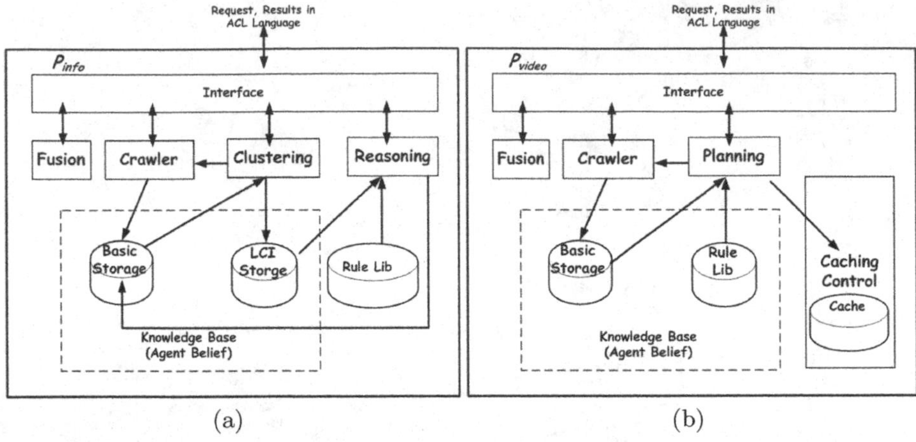

Fig. 3. Architecture of $\mathcal{P}_{info}$ and $\mathcal{P}_{video}$ Agent

rule matching algorithm, into the system. The inter-cluster reasoning rule can be simply designed as

– if $S_i.Nc_e.Ratio_e \geq \theta$, then $\mathcal{A}$.add(S_i).

To be noted that, the θ refers to the *reference threshold* about the remaining ratio of total storage space of cluster S_i's exemplar Nc_e. The operator "$x.y$" means a progressive function access from x to y.

While for intra-cluster reasoning, the reasoning rules are a bit more complex. More specially, for migrating those $\mathcal{E}$ caches to $\mathcal{F}$ caches, we have to consider the relationships between the $|\mathcal{E}|$ and $|\mathcal{F}|$, in which the symbol $|\bullet|$ represent the element number of a set. The $|\mathcal{F}|$ can be calculated by the formula $\sum |\mathcal{A}_i|$. Then the $|\mathcal{E}|$ can be calculated by the formula $\sum_{j=1}^{K} |S_j| - |\mathcal{F}|$. Therefore, the intra-cluster cache determination rules can be simplified and described as following:

– if the $|\mathcal{E}| = |\mathcal{F}|$, then C_{flag} will be set as *normal*;
– if the $|\mathcal{E}| > |\mathcal{F}|$, then C_{flag} will be set as *reuse*;
– if the $|\mathcal{E}| < |\mathcal{F}|$, then C_{flag} will be set as *refine*.

The $C_{flag} = $ '$normal$' means that the caching migration only takes a one-to-one mapping, the $C_{flag} = $ '$reuse$' means that at least one cache will be responsible for taking charge of two or more streams from other caches, and the $C_{flag} = $ '$refine$' means that the caching migration needs do a *top-k*$(k = |\mathcal{E}|)$ mapping.

3.3 Reasoning-Based Stream Transmission Planning

Based on the inter-cluster and intra-cluster reasoning, we can develop the corresponding stream transmission planning algorithm (See Alg. 1) for caching control.

Algorithm 1. Reasoning-based Stream Transmission Planning Algorithm for Caching Control

Require: LN, the reference caching threshold $LN.\theta$, candidate cluster set $LN.\mathcal{A}$, intra-cluster reasoning flag $LN.C_{flag}$, threshold adjustment $LN.\delta$.
Ensure: Give the planning for new transmission node set for each LN-level caches, i.e. $LN.\mathcal{F} = \{Nc'_1, Nc'_2, ..., Nc'_{|LN.\mathcal{F}|}\}$.

1: **initialization** : $LN.\mathcal{F} = null, LN.\mathcal{A} = null, LN.C_{flag} = null$;
2: **while** each LN **do**
3:   *generate* $LN.\mathcal{A}$ *based on inter-cluster reasoning in* $LN.\theta$
4:   get the $LN.\mathcal{E}$ and $LN.\mathcal{F}$ and compute their size
5:   *start to compute and get* $LN.C_{flag}$ *based on intra-cluster reasoning*
6:   **if** $LN.C_{flag} = \text{'}normal\text{'}$ **then**
7:     $LN.\mathcal{F}$.add($LN.\mathcal{A}.S_i$)
8:   **end if**
9:   **if** $LN.C_{flag} = \text{'}reuse\text{'}$ **then**
10:     $LN.\mathcal{F}$.add($LN.\mathcal{A}.S_i$)
11:     $LN.\theta <= LN.\theta - LN.\delta$
12:   **end if**
13:   **if** $LN.C_{flag} = \text{'}refine\text{'}$ **then**
14:     $LN.\mathcal{F}.add(TopRanking(\mathcal{A})_l)$
15:   **end if**
16:   output $LN.\mathcal{F}$
17: **end while**

In Alg. 1, in line 14, the $TopRanking(\mathcal{A})$ is a function to rank the set $\mathcal{A}$ in a descending order, while $TopRanking(\mathcal{A})_l$ represents the l-th element of this ranked set. Please be noted that, the line 11 ($LN.\theta <= LN.\theta - LN.\delta$) will be responsible for the adjustment of the threshold. The "*reuse*" status means our approach has filtered too many caches and only few caches are left for caching control, which may decrease the controlling flexibility and caching performance, therefore we should reduce the limit by the formula $LN.\theta <= LN.\theta - LN.\delta$. Through this adjustment, the inter-cluster reasoning, the intra-clustering reasoning and the dissemination algorithm would be dynamical running. As for the parameter setting, we will give corresponding suggestions in the experiment part. Based on the planning results, we can get the $LN.\mathcal{F}$, to which the filtered caches $LN.\mathcal{E}$ should be mapped and corresponding video stream can be migrated as well. In other words, the caches in $LN.\mathcal{F}$ should provide available *IP pool* to replace those in $LN.\mathcal{E}$.

4 Experimental Evaluations

In order to evaluate the performance of the proposed *AC3Caching*, we have constructed simulations using the network simulator NS_2 (http://www.isi.edu/nsnam/ns/). The simulated surveillance network topology is shown in Fig. 2, in which we totally deploy 20 IP camera(*IPC*) nodes. Each *IPC* node can read uncompressed video sequences, then encode and send data to the network in order to imitate the video stream generated by HD surveillance cameras.

Due to high code rate and multi-channel link sharing, 50MB size of the cache queue will be set in the sender module of each *IPC*. In addition, the video stream is firstly sent to a local data management server (*LDMS_1*,. . . ,*LDMS_8*) which has a bigger cache storage (1GB), and then be transmitted to the corresponding proxy data server (*PDS_1*,. . . ,*PDS_4*) which has 10GB of storage space. Finally, the video stream will be in read form *PDS*, decoded and displayed by the receiver. Additionally, in order to realize the proposed AC3Caching model, corresponding to three roles of agent mentioned in Sec. 3.2, three agent servers, i.e., *CAS*, *TAS*, *MAS*, need to access the simulated network. For each role of agent (e.g. *CA*), the related three types of agent(*info*, *video* and *control*) are jointly realized on the same server (e.g. *CAS*). The link between the sender side router and the receiver side router is set to be with 25ms delay, and the others are set to be with 5ms delay to simulate the real surveillance application.

In the simulations, we mainly evaluate the end-to-end *transmission latency.* We compare the results achieved when the system without and with *AC3Caching* running. In the first system, each *IPC* will be manually configured to transmit along with pre-specified *LDMS* and *PDS*. While in the second system, our dynamic transmission will be applied. Furthermore, within the evaluation about *transmission latency*, we will also evaluate the parameters including clustering frequency f, the threshold θ and corresponding adjustment δ, so as to find out how they will affect the performance of the proposed approach and to give corresponding parameter setting suggestions.

We transmitted a set of video sequences named *controlled-burn* (See Fig. 4(a)) from *IPC* to *PDS*. The sample sequences we used are unified captured with frame size of $1280 * 720$ pixels. Then we randomly select 15 *IPC*s to send the video with 15*fps*, and the rest *IPC*s with 30*fps*. As a result, different transmission pressure on servers can be generated. We will run each end-to-end transmission process at a fixed 20 minutes and obtain the the average frame delay. To be noted that, the delay involved in our approach only includes the transmission delay through the core network, and the encoding and decoding delay are excluded.

As shown in Fig. 4(b), the average delays with and without *AC3Caching* are computed respectively. We can see that, at the beginning of transmission, the one with *AC3Caching* will result in a larger delay. This is because that, at first, the differences among clusters may be relative unobvious. Hence, under such condition, unfortunately, the frequent changing of transmission path may increase the unexpected delay. As time goes by, transmission delay without *AC3Caching* will increase gradually. However, the agents in *AC3Caching* approach could perceive different storage pressures between caches and provide a new path through

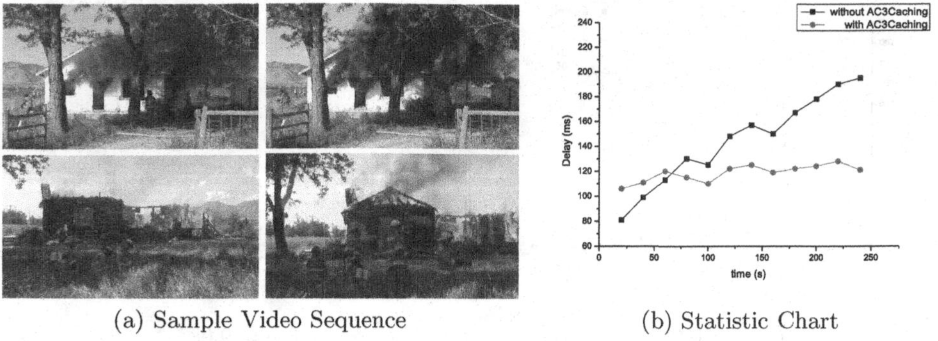

(a) Sample Video Sequence (b) Statistic Chart

Fig. 4. Video Surveillance System Running Efficiency with and without *AC3Caching*

their reasoning and cooperation. Finally, the proposed approach will take clear effect on transmission optimization and control the delay at a relatively stable value.

We make the assumption that, once the *IPC*s have generated N MB streams with time TP_N, the re-clustering will be triggered again. So the clustering frequency f can be calculated by the formula $f = 1/TP_N$. Initially, the threshold θ will be set to 40%, 50%, 60% and 70% respectively, and the adjustment δ will be set to 0.05, 0.08, 0.10 and 0.15. As Fig. 5(a) shown, under the same threshold θ(e.g. 40%), a larger f will generate a smaller average transmission delay. That's because that, when a heavy traffic load appears at certain links, our approach could promptly re-cluster and make corresponding transmission, which can decrease the transmission delay of video frames. Moreover, we can see that, an inflection point emerges when $f = 1/TP_{600}$ in Fig. 5(a). Taking the line with $\delta = 0.05$, $\theta = 40\%$ as an example, at this inflection point, the average delay may drop to 88ms. After this inflection point, even though taking a bigger re-clustering frequency $f = 1/TP_{100}$, the *delay* only can be reduced from 88ms to 80ms. This shows that in this condition after the inflection point, the re-clustering could take more computing cost and bring less benefit for delay decreasing. We can also see that when θ increases up to 70%, the delay becomes bigger than the one at $\theta = 40\%$. Through comparing the Fig. 4(a) with other three sub-figures, we can find that, when the adjustment δ is set to different values, corresponding inflection points will be different. In general, bigger δ will lead to earlier inflection point appearing.

To wrap up, a larger re-clustering frequency f will generate a smaller average transmission delay, and for same f , bigger δ will lead to earlier inflection point appearing. This shows that larger f and the smaller δ, in other word, more frequently re-clustering and smaller pace of adjustment for the reasoning-based planning process by agents, could achieve better performance for the proposed approach.

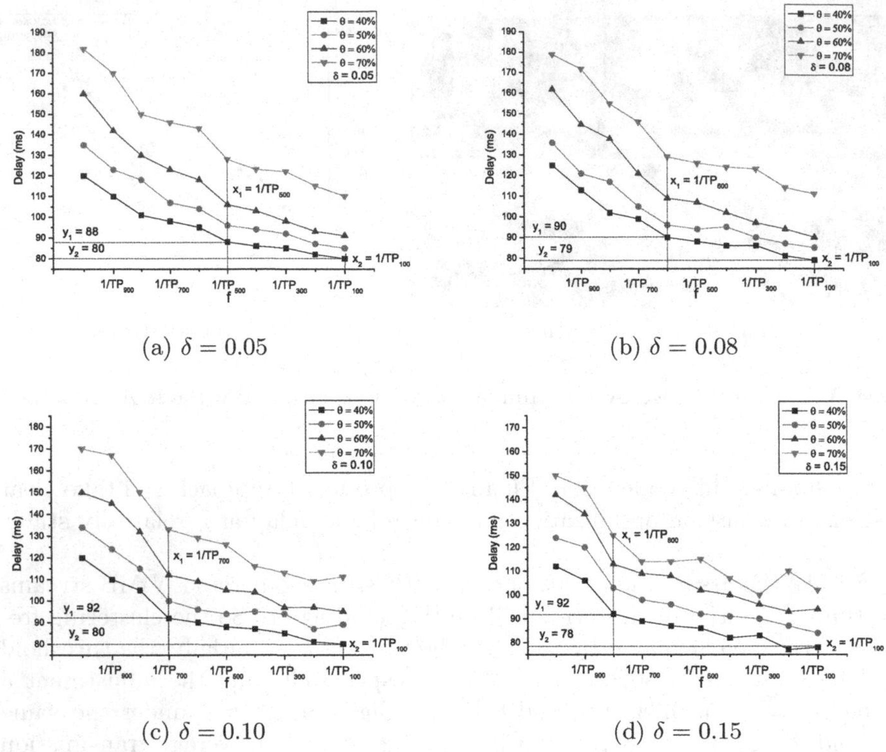

(a) $\delta = 0.05$ (b) $\delta = 0.08$

(c) $\delta = 0.10$ (d) $\delta = 0.15$

Fig. 5. Transmission Delay under Different f and θ with Increasing δ

5 Conclusions

Multi-agent collaboration-based caching control is a new paradigm for HD video surveillance. By introducing the multi-agent in cache management, resources related to knowledge flow (e.g. the *info flow*, the *video flow* and the *control flow*) can be automatically and dynamically managed by agents. Furthermore, the cooperative characteristics make the agents can collaborate with each other to accomplish the complex task (e.g caching control and video transmission) through the reasoning and planning ability.

As an extension of our continuous effort towards intelligent agent for service management [12] [20], this paper proposed the *AC3Caching* model together with two significant contributions:

1. Through providing a systematic collaboration mechanism by exploiting the *cluster characteristic* of cache storage, we extend a traditional agent model with the support of *AP clustering* and *caching optimization*.
2. Two reasoning methods,i.e., intra-cluster and inter-cluster reasonings are proposed respectively, so that the video transmission planning will be carried out effectively.

The evaluation for end-to-end *transmission latency* in HD video surveillance is discussed. As for how to set the parameters including clustering frequency f, the threshold θ and corresponding adjustment δ in our approach, we also give corresponding analysis and suggestions. Although our empirical results indicate that larger f and smaller δ could achieve better performance, this parameter setting will result in heavy clustering-related cost f. Hence, the adaptive dynamical f determination need further study for improving this approach.

References

1. Cao, L.: Data mining and multi-agent integration. Springer-Verlag New York Inc. (2009)
2. Cao, L., Gorodetsky, V., Mitkas, P.A.: Agent mining: The synergy of agents and data mining. IEEE Intelligent Systems 24(3), 64–72 (2009)
3. Cao, L., Weiss, G., Yu, P.S.: A brief introduction to agent mining. In: Autonomous Agents and Multi-Agent Systems, pp. 1–6 (2012)
4. Cheung, G., Tan, W., Yoshimura, T.: Rate-distortion optimized application-level retransmission using streaming agent for video streaming over 3g wireless network. In: Proceedings of the 2002 International Conference on Image Processing, vol. 1, pp. I–529. IEEE (2002)
5. DeLisi, C.: Theory of clustering of cell surface receptors by ligands of arbitrary valence: dependence of dose response patterns on a coarse cluster characteristic. Mathematical Biosciences 52(3-4), 159–184 (1980)
6. Frey, B.J., Dueck, D.: Clustering by passing messages between data points. Science 315(5814), 972 (2007)
7. Karp, R.M., Rabin, M.O.: Efficient randomized pattern-matching algorithms. IBM Journal of Research and Development 31(2), 249–260 (2010)
8. Krishnaswamy, S., Zaslavsky, A., Loke, S.W.: An architecture to support distributed data mining services in e-commerce environments. In: Wecwis, p. 239. IEEE Computer Society (2000)
9. Lawton, G.: Working today on tomorrow's storage technology. Computer 39(12), 19–22 (2006)
10. Lee, U., Magistretti, E., Gerla, M., Bellavista, P., Liò, P., Lee, K.W.: Bio-inspired multi-agent collaboration for urban monitoring applications. In: Bio-Inspired Computing and Communication, pp. 204–216 (2008)
11. Li, C., Gao, Y.: Agent-based pattern mining of discredited activities in public services. In: 2006 IEEE/WIC/ACM International Conference on Web Intelligence and Intelligent Agent Technology Workshops, WI-IAT 2006 Workshops, pp. 15–18. IEEE (2007)
12. Niu, W., Li, G., Zhao, Z., Tang, H., Shi, Z.: Carsa: A context-aware reasoning-based service agent model for ai planning of web service composition. Journal of Network and Computer Applications (2011)
13. Ponnusamy, S.P., Kathikeyan, E.: Hp proxy: Hot-point proxy caching with multivariate sectoring for multimedia streaming. European Journal of Scientific Research 68(1), 21–35 (2012)
14. Ramesh, S., Rhee, I., Guo, K.: Multicast with cache (mcache): An adaptive zero-delay video-on-demand service. IEEE Transactions on Circuits and Systems for Video Technology 11(3), 440–456 (2001)

15. Rejaie, R., Kangasharju, J.: Mocha: A quality adaptive multimedia proxy cache for internet streaming. In: Proceedings of the 11th International Workshop on Network and Operating Systems Support for Digital Audio and Video, pp. 3–10. ACM (2001)
16. Shen, B., Lee, S.J., Basu, S.: Caching strategies in transcoding-enabled proxy systems for streaming media distribution networks. IEEE Transactions on Multimedia 6(2), 375–386 (2004)
17. Tran, D.A., Hua, K.A., Sheu, S.: A new caching architecture for efficient video-on-demand services on the internet. In: Proceedings of the 2003 Symposium on Applications and the Internet, pp. 172–181. IEEE (2003)
18. Wang, B., Sen, S., Adler, M., Towsley, D.: Optimal proxy cache allocation for efficient streaming media distribution. In: Proceedings of the IEEE Twenty-First Annual Joint Conference of the IEEE Computer and Communications Societies, INFOCOM 2002, vol. 3, pp. 1726–1735. IEEE (2002)
19. Wang, C.H., Chang, R.I., Ho, J.M.: Collaborative video surveillance for distributed visual data mining of potential risk and crime detection. In: Surveillance Technologies and Early Warning Systems: Data Mining Applications for Risk Detection, p. 194 (2010)
20. Wang, X., Niu, W., Li, G., Yang, X., Shi, Z.: Mining Frequent Agent Action Patterns for Effective Multi-agent-Based Web Service Composition. In: Cao, L., Bazzan, A.L.C., Symeonidis, A.L., Gorodetsky, V.I., Weiss, G., Yu, P.S. (eds.) ADMI 2011. LNCS, vol. 7103, pp. 211–227. Springer, Heidelberg (2012)

System Modeling of a Smart-Home Healthy Lifestyle Assistant

Xinhua Zhu[1], Yaxin Yu[2], Yuming Ou[1], Dan Luo[1], Chengqi Zhang[1], and Jiahang Chen[1]

[1] AAI, QCIS, FEIT, University of Technology, Sydney, Australia
{Xinhua.Zhu,Yuming.Ou,Dan.Luo,Chengqi.Zhang,Jiahang.Chen}@uts.edu.au
[2] College of Information Science and Engineering, Northeastern University, China
Yuyx@mail.neu.edu.cn

Abstract. A system modeling is presented for a Smart-home Healthy Lifestyle Assistant System (SHLAS), covering healthy lifestyle promotion by intelligently collecting and analyzing context information, executing control instruction and suggesting health plans for users. SHLAS is Multi-agent based. Each agent has three levels: the Goal Layer has business rules for representing agent goals; the Strategy Layer provides technical rules and processes for guiding how the agent reacts to events; the Component Layer is made up of components, some components are called by technical rules and processes in the Strategy Layer, some others are used for communicating with third party systems. This agent framework enables the customizability of agents in SHLAS. We also introduce an Ontology-based domain knowledge and context model to capture and represent the agents, and agent behavior which provides agents with reasoning ability. SHLAS helps users with healthy lifestyle promotion by tracking and analyzing their behaviors, and recommending health plans. The paper closes with an empirical evaluation of the approach from the point of view of customizability.

Keywords: Multi-agent, Agent behavior analysis, Customizability, Planning.

1 Introduction

The smart home is always a hot topic in both academe and industry. Science fiction has imagined an idealized vision of a fully integrated smart home, where all the operations of a house can be efficiently controlled by a central application. A good smart home system should be an excellent assistant to home life, able to collect and analyze a range of information from people's daily lives and use it to optimize their living environment. It should be able to flexibly control home appliances, monitor people's health status, advise of any abnormalities, develop personalized health programs for each family member, and push users to execute such plans. Using smart home systems, people would have ease of access to information and services that will improve health and quality of life.

L. Cao et al.: ADMI 2012, LNAI 7607, pp. 65–78, 2013.

A significant feature of a smart home is intelligence. According to Wikipedia, Intelligence has been defined in different ways, including, but not limited to, the ability to exercise abstract thought, learning, understanding, self-awareness, memory, reasoning, planning, communication, emotional knowledge, and problem solving[1]. This is not the case with current smart home products. Most existing smart home systems are not intelligent enough. The main issues include:

1) Oversimplified processing logic. The current smart home systems such as MavHome [8] rely heavily on sensor data acquisition and the passive acceptance of clear control commands, and lack intelligent and complex human-computer interaction and reasoning ability. It is hard for users to convey their intentions in a natural way through speech, text or gesture.

2) Poor flexibility and scalability. Different families may have different requirements and they may change their requirements frequently. Moreover, devices may be added and removed during runtime and they may fail due to connectivity problems. As a result, there is a need for an open complex agile architecture to deal with all these circumstances.

3) Insufficient focus on healthy lifestyle promotion. In the final analysis, people's health mostly depends on their daily behaviors. Early awareness of inappropriate habits could help to improve health and prevent disease. However, to the best of our knowledge, there is limited research studying the smart home from the perspective of healthy lifestyle promotion.

The paper proposes a Smart-home Healthy Lifestyle Assistant System (SHLAS). In respect of the first problem above, oversimplified processing logic, SHLAS draws on the Cyber-physical system (CPS) [2] theory. CPS is a multi-dimensional complex intelligent system integrated with computing, networking and the physical environment. Real-time perception, dynamic control, and information services of large-scale engineering CPS can be implemented by use of 3C (Computation, Communication, Control) technologies. In CPS, a key feature is context-aware computing which focuses on the collection, modeling, and intelligent processing of context information [1].Our SHLAS makes full use of context-aware computing. At the information collection stage, similar to the induction of the human sensory organs to the objective world, SHLAS obtains context information related to people and equipment through human-computer interaction and a Wireless Sensor Network (WSN). At the modeling stage, the collected multi-source heterogeneous information is transformed, represented and stored effectively. At the intelligent processing stage, useful and meaningful know-ledge is discovered from the collected information so that SHLAS can understand users' behaviors and intentions. Finally context-aware services are provided.

In terms of the second problem, poor flexibility and scalability, SHLAS introduces an open, complex, agile multi-agent architecture model. Agent is the basic execution unit in SHLAS. Agents are grouped in accordance with their goals and

[1] `http://en.wikipedia.org/wiki/Intelligence/`

[2] `http://en.wikipedia.org/wiki/Cyber-physical_system/`

events are used in agent communication. Each agent is modeled carefully to regulate its behavior and interface. An agent model can be divided into three layers: the Component Layer, Strategy Layer and Goal Layer. The Component Layer provides a variety of internal function components called by the Strategy Layer and external interface components for sensor data collection, device control and software control; the Strategy Layer defines basic agent behavior logics through processes and rules; and the Goal Layer defines high level business rules written in natural language to represent agents' goals. Agent behaviors are highly customizable by updating rules and processes, which ensures flexibility on the basis of commonality.

To address third problem of healthy lifestyle promotion, a Health Promotion Procedure Model (HPPM) is introduced which combines health domain knowledge, context, intelligent planning [2] and behavior informatics [3] [4]. HPPM is an iterative four-step management model: planning, reminder, behavior monitoring and performance evaluation. Four important factors are considered in the planning stage: health promotion knowledge; historical behavior analysis; current status; and health goals.

The paper is organized as follows. In Section 2, the concept of a smart environment, typical projects and related tools are discussed. Section 1 introduces the architecture of SHLAS. Section 4 proposes the ontology-based domain knowledge and context modeling. We illustrate two typical application scenarios in Section 5. Several key points in implementation are discussed in Section 6. We conclude the paper in Section 7.

2 Related Work

A smart environment should be able to acquire and apply knowledge about its inhabitants and their surroundings in order to adapt to the inhabitants and meet the goals of comfort and efficiency [5]. It has further been stated that a smart environment can reduce the amount of interaction required by inhabitants, reduce energy consumption and limit other potential waste, and provide a mechanism for ensuring the health and safety of the environment's occupants [6].

EasyLiving [7] at Microsoft Research supports smart environments through the dynamic interconnection of a variety of devices. This middle-ware supports mechanisms such as inter-system communication, location tracking for objects and people, and visual perception. One issue confronting the EasyLiving system is that it only focuses on room control and lacks residential health management.

The MavHome Smart Home project [8] is a multi-disciplinary research project at Washington State University and the University of Texas at Arlington which focuses on the creation of an intelligent home environment. It views the smart home as an intelligent agent that perceives its environment through the use of sensors and can act upon the environment through the use of actuators. MavHome provides many excellent features such as resident behavior analysis and action prediction, but makes no effort to improve flexibility; for instance, how to adjust agent behavior easily.

Rule based tools can provide a smart home system with powerful flexibility. A Business Logic Integration Platform called Drools has been introduced which provides a unified and integrated platform for Rules, Workflow and Event Processing. It has been designed from the ground up so that each aspect is first class, with no compromises [3]. The Drools Planner is used to optimize automated planning by combining search algorithms with the power of the Drools rule engine[4]. All Drools components are used in SHLAS for agent behavior representation and agent communication.

Data Mining and Multi-agent Integration [11] [12] [13] presents cutting-edge research, applications and solutions in data mining. This will improve smart home system on both performance and intelligence.

3 Architecture of SHLAS

Agent is the basic execution unit in SHLAS. As shown in Fig. 1, there are three groups of agents in SHLAS: the Core Agent Group, Interface Agent Group and Service Agent Group. We draw on the agent service abstract model [9] to develop a theoretical model of a smart home agent.

A smart home agent is represented by attributes such as name, type, locator, owner, roles, behavior, protocol, address, event, input variables, pre-conditions, output variables, post-conditions and exception handling.

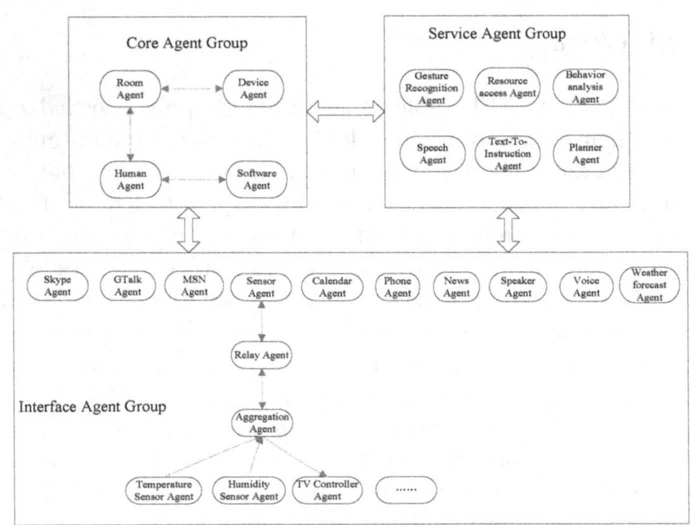

Fig. 1. SHLAS Architecture

[3] http://www.jboss.org/drools/

[4] http://docs.jboss.org/drools/release/5.3.0.Beta1/drools-planner-docs/html_single/#d0e26/

$$<Smart-HomeAgent>::=f(Name;Type;Behavior;Protocol;Locators;Owner;Roles;Address;Event;InputVariables;Preconditions;OutputVariables;Postconditions;Exception). \quad (1)$$

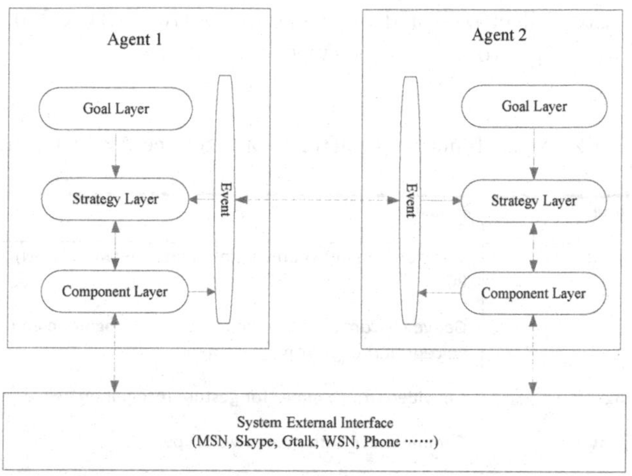

Fig. 2. Smart-Home Agent Model

Behavior is one of the most important elements in the Smart-Home Agent Model. As shown in Fig. 2, behaviors are redistributed into three layers of agent architecture, namely, Goal Layer, Strategy Layer and Component Layer.

Table 1. Agent functional patterns of core Agent Group

Agent name	Behavior
Human Agent	Represent member states, determine control privilege, and analyze behavior patterns.
Room Agent	Represent room states, create control-instruction event to improve room environment or alert user if any abnormality.
Device Agent	Represent device states and control device actions.
Software Agent	Represent software states and control software actions.

- In Goal Layer, high level business rules written in natural language are used to express agent goals. It enables even end users to adjust agent behavior easily by updating the business rules.

– Strategy Layer defines the elements such as technical rules and processes to represent agent behavior. Parts of the technical rules are converted from business rules in the Goal Layer. Others are defined by developers. All the processes and rules are easy to update through a GUI tool. Processes and rules can be triggered by events for agent communication.
– Component Layer provides a variety of basic components whose logic will not be changed frequently. Some components are called by processes and rules in the Strategy Layer. Others interact with system external interfaces, for instance, to collect sensor data, to send control instructions, to interact with MSN or Skype, to call web service, and so on.

Table 2. Agent functional patterns of Interface Agent Group

Agent name	Behavior
Sensor Agent	Transfer data between wireless sensor network (WSN) and PC
Voice Agent	Get voice from microphone for speech recognition and voiceprint recognition.
Camera Agent	Get video from camera for gesture recognition.
Skype Agent	Communicate with people via Skype.
GTalk Agent	Communicate with people via GTalk.
MSN Agent	Communicate with people via MSN.
Phone Agent	Communicate with people via Phone.
Calendar Agent	Get and set Calendar Item.
News Agent	Get news from news RSS websites.
Weather Forecast Agent	Get weather forecast information from weather forecast RSS websites.
Speaker Agent	Play speech.
Relay Agent	Relay transferred data.
Aggregation Agent	Connect and transfer data between Terminal Sensor Agent and Relay Agent.
Terminal Sensor Agent	Collect sensor data and control devices.

Family members, rooms, domestic devices and software applications are the major management objects in SHLAS. As shown in Table 1, the Core Agent Group consists of Human Agent, Room Agent, Domestic Device Agent and Domestic Software Agent.

The Interface Agent Group deals with data exchange between SHLAS and third party systems. As shown in Table 2, the Interface Agent Group includes

Table 3. Agent functional patterns of Service Agent Group

Agent name	Behavior
Gesture Recognition Agent	Map gesture to instruction.
Resource Access Agent	Get context resource.
Behavior Analysis Agent	Analyze people behavior.
Speech Agent	Map speech to text.
Text-To-Instruction Agent	Map text to instruction.
Planner Agent	Make health plan.

Sensor Agent, Speech Agent, Camera Agent, Skype Agent, GTalk Agent, MSN Agent, Phone Agent, Calendar Agent, and more.

As shown in Table 3, the last group is the Service Agent Group, which comprises Gesture Recognition Agent, Resource Access Agent, Behavior Analysis Agent, Speech Agent, Text-To-Instruction Agent and Planner Agent. These agents provide passive services for other agents to call.

4 Ontology Based Domain Knowledge and Context Modeling

Domain knowledge and context are significant resources in SHLAS. As shown in Fig. 3, we overview the Ontology Based Domain Knowledge and Context Model.

Domain knowledge is composed of terms and their relationships. For instance, in the domestic domain, there are open spaces (such as the balcony and garden) and rooms (such as the bedroom, toilet, living room, etc). A living room may contain devices such as TV, air conditioner, washing machine, etc. A TV has instructions such as turn on/off, channel tune and volume tune. An authorized family member may execute particular instructions. Fig. 4 shows parts of the domain knowledge; users can construct a domain knowledge model by dragging and dropping a word from the left column to the right column, naming it, and linking it to other terms.

Some special terms can be further defined as object templates represented in an xml scheme (XSD file). Fig. 5 shows a TV scheme template example. The template describes the object name, instructions and corresponding instruction execution method, and so on.

Context represents the instance of domain knowledge. For example, Family A has a master bedroom and a secondary bedroom; in master bedroom there are a SHARP TV and a TOSHIBA Air Conditioner, etc. Context information can be easily managed. As shown in Fig. 6, when an object type is selected, such as TV, information in the corresponding XSD scheme will be extracted and displayed on the GUI, which is very convenient for users because they do not need to spend much time defining all the objects themselves.

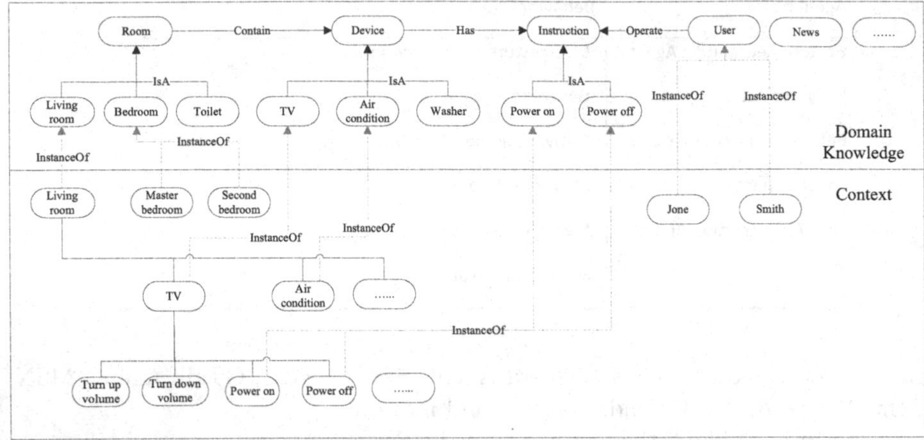

Fig. 3. Domain knowledge and context model

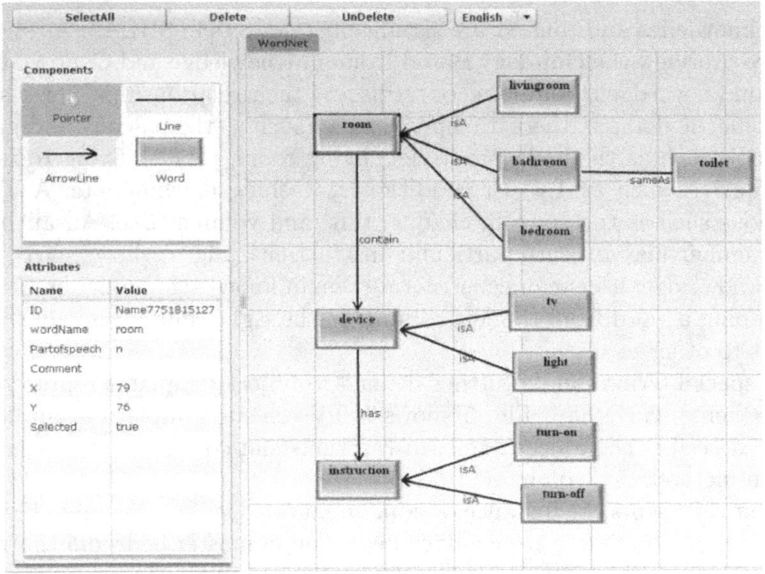

Fig. 4. Domain Knowledge Management

```xml
<?xml version="1.0" encoding="UTF-8"?>
<xsd:schema xmlns="SHLAS.Schema"
    xmlns:xsd="http://www.w3.org/2001/XMLSchema"
    targetNamespace="SHLAS.Schema"
    elementFormDefault="qualified"
    attributeFormDefault="unqualified"
>
<xsd:element name="TeleVision">
    <xsd:complexType>
        <xsd:sequence>
            <xsd:element name="Instructions">
                <xsd:complexType>
                    <xsd:all>
                        <xsd:element name="TurnOn">
                            <xsd:complexType>
                                <xsd:sequence>
                                    <xsd:element name="JavaBeanName" type="xsd:string" fixed="SHLAS.Instruction.InferedSend"></xsd:element>
                                        <xsd:element name="Parameters">
                                            <xsd:complexType>
                                                <xsd:sequence>
                                                    <xsd:element name="Port">
                                                        <xsd:complexType>
                                                            <xsd:attribute name="value">
                                                                <xsd:simpleType>
                                                                    <xsd:restriction base="xsd:integer">
                                                                    </xsd:restriction>
                                                                </xsd:simpleType>
                                                            </xsd:attribute>
                                                        </xsd:complexType>
                                                    </xsd:element>
                                                </xsd:sequence>
                                            </xsd:complexType>
                                        </xsd:element>
                                        <xsd:element name="Roles">
                                            <xsd:complexType>
                                                <xsd:attribute name="Type" default="Adult">
                                                </xsd:attribute>
                                            </xsd:complexType>
                                        </xsd:element>
                                </xsd:sequence>
                            </xsd:complexType>
                        </xsd:element>
```

Fig. 5. XSD template of TV

Domain knowledge and context are used in reasoning. Details are described in Section 6.

5 Scenario Analysis

In this section, two typical scenarios and the corresponding agent behaviors are introduced to illustrate how SHLAS works. Generally speaking, people's health includes physical health, mental health, social relations health, and so on. Whether or not their daily lifestyle is rational will directly affect a person's health. Through behavioral analysis, habits can be acknowledged and personalized services can be recommended to help the member achieve a healthy lifestyle. Scenario 1 is an example of this.

Scenario 1: James worked hard all week. At the weekend, he slept until noon. When he woke, he said, "Open the curtain". The system knew that he was in the bedroom, so the cur-tain in the bedroom was opened. The system recognized that James got up later than usual, so he was reminded: "You got up a little later than usual; please take care of your sleep quality". James always bathed and listened to music after getting up, so the shower was powered on automatically and he was asked: "Would you like to listen to music?" If the answer is "Yes", the system collects and plays a music list based on his historical preference.

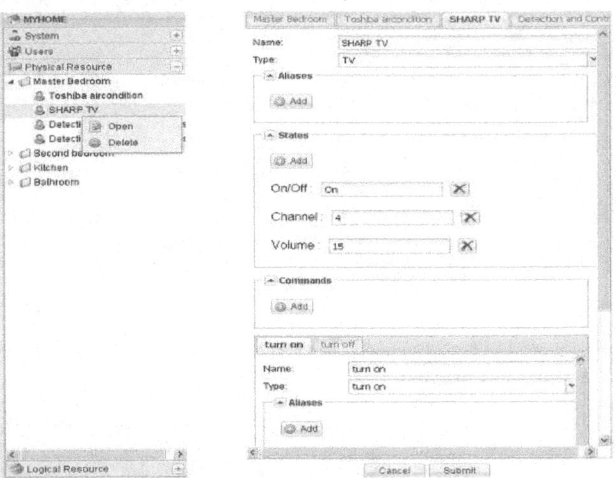

Fig. 6. Context information management

Fig. 7 shows the agent behavior diagram in Scenario 1. The figure omits the message passing procedure from Sensor Agent to Terminal Sensor Agent via Relay Agent and Aggregation Agent.

Scenario 1 is an application of behavior pattern analysis and personalized recommendation. To realize the scenes, SHLAS needs access to a variety of contexts, such as the basic states of the family members, behavioral preferences and environment information. To obtain behavioral preferences, SHLAS uses the FP-growth algorithm [10] to analyze the accumulated historical behavior data regularly in advance, and then stores the identified behavior patterns. Based on these behavior patterns, SHLAS determines the impact factors of health and recommends the most appropriate activities to family members.

It is rarely easy to achieve a long-term healthy state without any expertise or professional guidance. Diet and fitness rules are stored in SHLAS. Professional health plans can be established based on these rules if there are health requirements. Scenario 2 is an example.

Scenario 2: James stood on a weighing scale. The system found that he weighed 2kg more than last month and was 1kg more than standard weight, so James was told: "You are 1kg overweight. Would you like a health plan?" If the answer is "yes", the system collects information about his current status, analyzes his historical behaviors and goals, and generates a three-month plan. The plan is generated into Google Calendar. James can make modifications through any web browser. In the following three months, James will be reminded by speaker or email to execute the planning items wherever he is at home or in the office. James's behavior will be monitored and his ongoing performance will be evaluated for plan refinement.

Fig. 8 shows the sequence diagram of generating a health plan.

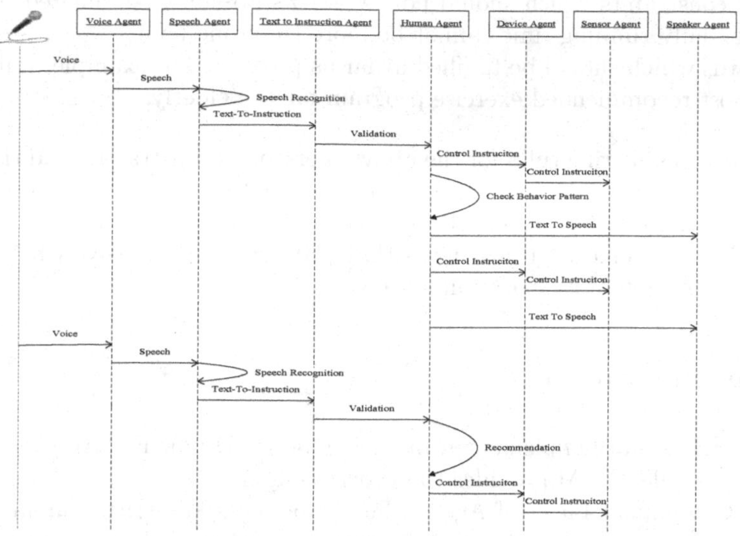

Fig. 7. Agent behavior in Scenario 1

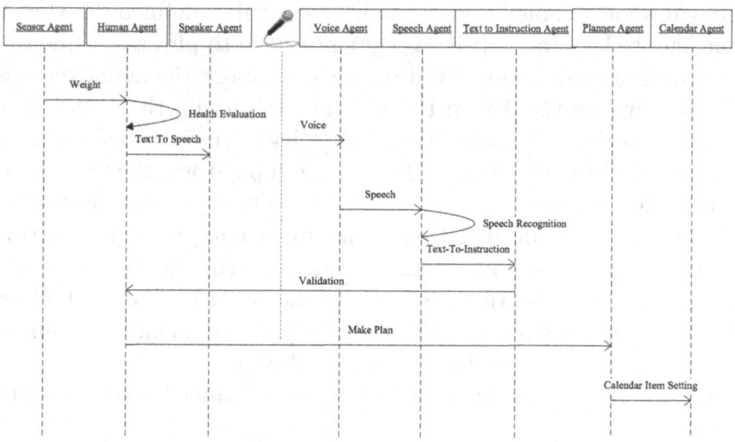

Fig. 8. Health plan generation in Scenario 2

SHLAS provides users with a complete Health Promotion Procedure Model (HPPM) which includes planning, reminding, monitoring, evaluating and plan refining. Scenario 2 is an application scenario of HPPM. Planning is based on the rules related to each family member's behavior patterns, current health status, home environment, expectations for health, and so on. The following constraints in the rules are especially important:

– Hard constraint which must not be broken. For example, only one exercise program for each member can be in progress at any given time.

– Weak constraints which should not, as far as possible, be violated. For example, daily running time cannot be more than one hour.
– Rewards which should be fulfilled as far as possible. For example, walking is the most recommended exercise program for the elderly.

SHLAS develops scoring rules for the above-mentioned constraints, and the rules are used to score a number of health plans generated automatically. One or more sets of plans with higher scores will be used by family members.

From the two scenarios, it is obvious that SHLAS comprehensively helps users with the promotion of a healthy lifestyle.

6 Implementation

Agents in SHLAS are implemented in Java Thread. Drools Expert provides the rule engine, while JBPM provides the process engine.

In the Component Layer of Agent, the components are written in JavaBean. A component interface specification is defined so that updating a component will not cause damage to other components.

In the Strategy Layer, agent behavior can be refactored by tuning the processes and rules of Drools. Fig. 9 shows two rules defined in the Text-To-Instruction Agent. The first rule is to segment text into phrases, while the second is to create an instruction event from phrases. Suppose the agent receives a text event: "Please turn on the TV in the master bedroom", the first rule will segment the text into three phrases, "turn on", "TV", "master bedroom", and will retrieve their types from the domain knowledge repository: the type of "turn on" is "instruction; the type of "TV" is "device"; and the type of "master bedroom" is "room". The second rule will then build up a completed instruction event. The event will trigger a rule in Human Agent to verify the privilege and finally will trigger a rule in Device Agent to execute the control action. All classes (e.g., PhraseEvent, InstructionEvent) and methods (e.g., segment) are defined in the Component Layer of the Text-To-Instruction Agent.

The rules can be refined by a developer or advanced user in runtime and enabled immediately.

In the Goal Layer, end users can modify business rules which are easier to under-stand than technical rules. The business rules in the Goal Layer will be translated into technical rules automatically. Fig. 10 shows a business rule defined in Room Agent; the first line is business rule, the second and the third lines contain the corresponding technical rule. During runtime, end users can only access the business rule. Once end users update "number" in the business rule to another value, the agent behavior will change correspondingly.

Compared to other smart home approaches, SHLAS is special in treating refinement as an optimization problem. Elements in the three layers are updatable and can be accessed by three levels of customers. A developer can refine the low level behavior of the agent by updating the source code of components. An advanced user can refine the medium level behavior of an agent by managing technical

```
rule "Segment text into phrases"
    when
        $text:TextEvent()
    then
        String[] phrases = segmentText($text.getContent());
        for(String phrase:phrases){
            PhraseEvent phraseEvent = new PhraseEvent();
            phraseEvent.setContent(phrase);
            phraseEvent.setType(getType(phrase));
            phraseEvent.setSource($text.getSource());
            insert(phraseEvent);
        }
end

rule "Create instruction event from phrases"
    when
        $room:PhraseEvent(type == PhraseType.ROOM)
        $device:PhraseEvent(type == PhraseType.DEVICE && source == $room.getSource
        $command:PhraseEvent(type == PhraseType.COMMAND && source == $room.getSourc
    then
        InstructionEvent instructionEvent = new InstructionEvent();
        instructionEvent.setLocation($room.getContent);
        instructionEvent.setDevice($device.getContent);
        instructionEvent.setCommand($command.getContent);
        instructionEvent.setSource($room.getSource());
        insert(instructionEvent);
end
```

Fig. 9. Rules in Text-To-Instruction Agent

processes and rules. A general end user can refine the high level behavior of the agent by managing business rules.

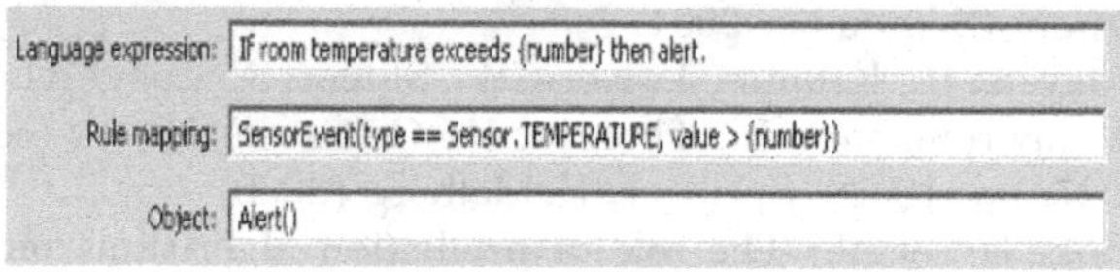

Fig. 10. Business rule and corresponding technical rule

7 Conclusion

This paper proposes a SHLAS system which can assist family members in the promotion of a healthy lifestyle by collecting and analyzing context information, executing control instruction and making health plans.

A multi-modal human-computer interaction style, healthy domain knowledge, and model-based and rule-based reasoning ability make SHLAS intelligent. Multi-agent architecture and a component-strategy-goal agent framework make SHLAS behavior much easier to customize than other smart home approaches. SHLAS helps users to promote a healthy lifestyle by making health plans and reminding the user to enact the plan. Health planning is based on health domain knowledge, collected context information and users' goals. Compared to other smart home approaches, the agent behavior in SHLAS is easier to refine for on-demand requirements due to the flexible multi-agent architecture used in this approach. In fact, this approach suggests a general architecture for managing SHLAS which can easily be extended to other intelligent spaces, such as hospital, office, school, and restaurant, as long as the agent behavior and domain knowledge are updated correspondingly. Our future work will focus on human computer interaction (e.g., emotion recognition) and agent behavior impact analysis (e.g., how the behaviors of Human Agent, Room Agent and Device Agent impact on one another).

References

1. Schilit, B., Adams, N., Want, R.: Context-aware computing applications. In: IEEE Workshop on Mobile Computing Systems and Applications, WMCSA 1994, Santa Cruz, CA, pp. 89–101 (1994)
2. Wilkins, D., Desjardins, M.: A call for knowledge-based planning. AI Magazine 22(1), 99–115 (2001)
3. Cao, L., Yu, P.S. (eds.): Behavior Computing: Modeling, Analysis, Mining and Decision. Springer, New York (2012)
4. Cao, L.: Behavior informatics and analytics: Let behavior talk. In: IEEE International Conference on Data Mining Workshops, ICDMW 2008, pp. 87–96 (2008)
5. Cook, D., Das, e.S.: Smart Environments: Technologies, Protocols and Applications. Wiley (2004)
6. Das, S.K., Cook, D.: Health monitoring in an agent-based smart home. In: Proceedings of the International Conference on Smart Homes and Health Telematics, ICOST, Singapore (September 2004)
7. Brumitt, B., Meyers, B., Krumm, J., Kern, A., Shafer, S.: EasyLiving: Technologies for Intelligent Environments. In: Thomas, P., Gellersen, H.-W. (eds.) HUC 2000. LNCS, vol. 1927, pp. 12–29. Springer, Heidelberg (2000)
8. Das, S.K., Cook, D., et al.: The role of prediction algorithms on the MavHome Smart Home Architectures. IEEE Wireless Communications (Special Issue on Smart Homes) 9(6), 77–84 (2002)
9. Cao, L., Zhang, C., Zhou, M.C.: Engineering open complex agent systems: A case study. IEEE Transactions on Systems, Man, and Cybernetics-Part C: Applications and Reviews 38(4), 483–496 (2008)
10. Han, J., Pei, J., Yin, Y., Mao, R.: Mining frequent patterns without candidate generation. Data Mining and Knowledge Discovery 8, 53–87 (2004)
11. Cao, L.: Data Mining and Multi-agent Integration (edited). Springer (2009)
12. Cao, L., Weiss, G., Yu, P.S.: A Brief Introduction to Agent Mining. Journal of Autonomous Agents and Multi-Agent Systems 25, 419–424 (2012)
13. Cao, L., Gorodetsky, V., Mitkas, P.A.: Agent Mining: The Synergy of Agents and Data Mining. IEEE Intelligent Systems 24(3), 64–72 (2009)

Part III

Data Mining for Agents

An Optimization Approach to Believable Behavior in Computer Games

Yifeng Zeng[1], Hua Mao[1], Fan Yang[2], and Jian Luo[2]

[1] Department of Computer Science, Aalborg University, Denmark
{yfzeng,huamao}@cs.aau.dk
[2] Department of Automation, Xiamen University, China
{yang,jianluo}@xmu.edu.cn

Abstract. Many artificial intelligence techniques have been developed to construct intelligent non-player characters (NPCs) in computer games. As games are gradually becoming an integral part of our life, they require human-like NPCs that shall exhibit believable behavior in the game-play. In this paper, we present an optimization approach to designing believable behavior models for NPCs. We quantify the notion of believability using a multi-objective function, and subsequently convert the achieving of believable behavior into one function optimization problem. We compute its analytical solutions and demonstrate the performance in a practical game.

1 Introduction

Designing intelligent non-player characters (NPCs) has been a focus of game designers and developers who often resort to sophisticated techniques in the area of artificial intelligence (AI). As expected, the resulting NPCs make smart, nearly optimal, decisions in a complex game world. For example, a tennis NPC may strongly attack human players in the virtual tennis game [16] and you may be impressed by tricky plans of *Frederick* and *Gandhi* in the *Civilization IV* [1]. Currently it is not rare that intelligent NPCs may defeat experienced human players. However, the NPCs' actions tend to appear artificial after some time of playing, and rule out any surprise that human opponents would provide. This has motivated a line of research on constructing believable NPCs in interactive games [4,6,20,21].

Much of the existing research takes the macro-perspective on designing believable behavior. For example, the *Soar* architecture provides a cognitive model to develop believable agents in computer games [9,19]. In parallel, the *ICARUS* framework facilitates the development of goal-directed agents in games [10]. The *Emotivector* model encodes an anticipatory mechanism for the believability enhancement on designing human-like characters [11]. While the mentioned research significantly drives the study on believable behavior, it requires much effort to integrate the associated frameworks or models into the routine design of NPCs in game productions. In this paper, we will adapt behavior trees [7,3]

L. Cao et al.: ADMI 2012, LNAI 7607, pp. 81–92, 2013.

- a new generation of script language for game design - to develop believable behavior models of NPCs.

The notion of believability has been studied in the fields of arts, psychology and computer science, for a couple of decades [2,13]. Linking to NPCs in computer games, believable behavior is not especially smart and is coupled with some unpredictability [14,17]. In other words, behaviors that are too intelligent will be rapidly categorized as being unreal, and that are very unpredictable may lead to the feeling of randomness. We need to make a proper tradeoff between the intelligence and randomness of the behavior. According to this spirit, the objective on designing believable behavior is to achieve the intelligence of NPCs' behavior and simultaneously maintain the diversity of their behavior. We will formulate the believability design as one multi-objective optimization problem and compute its optimal solutions if they exist.

We focus on the realization of believable behavior based on behavior trees. As behavior trees plan NPCs' actions in the game-play we may construct the optimal behavior through AI planning and learning techniques [15]. To make NPCs act intelligently, we let their behavior approach the optimal one. We use a probability-based distance measurement to quantify the intelligence of NPCs' behavior. Meanwhile, we use the information entropy [5] to measure the diversity of NPCs' behavior. The combination of these two measurements provides a quantitative approach to formulate the believability of NPCs' behavior. The formulation allows us to compute the best believability. Finally, we evaluate the believability design in both simulations and user tests, and demonstrate the practical utility of our techniques.

2 Background: Behavior Tree

Behavior tree is a graphical representation for structuring NPCs' behavior in a modular manner so that both designers and programmers can work together in the game development [7,3]. It starts with a *Root* node (denoted by a down triangle shape) and normally ends with an *Action* node (denoted by a circle shape) as its leaf. We show the other three types of basic nodes in Fig. 1. We refer the reader to [7,3] for more details on the representation of behavior trees.

A *Sequence* node has one or more *Action* nodes as its children, and is used whenever a sequence of actions have dependency upon one another. The *Sequence* executes the first action ($A1$), and if the execution returns success it continues the next one ($A2$) and so on. Failure of any action terminates the execution of the *Sequence* node. A *Selector* node contains a set of independent *Action* nodes, and may choose one of them for an execution. The *Selector* fails only if neither of its children nodes can be successfully executed. In general, game designers assign a probability distribution, $(p_{A_1},\cdots,p_{A_N})$, over the set of N actions. The *Selector* executes one of the actions according to its probability distribution. A *Decorator* node is inserted on the top of an action node or a subtree in order to provide additional functionalities to a generic behavior. For example, one type of *Decorator* can either limit the number of times that

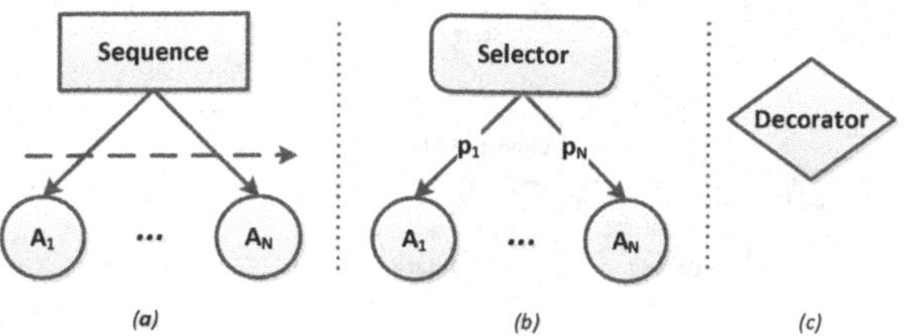

Fig. 1. (*a*) A *Sequence* node with N actions $(A_1, \cdots, A_N)$ where the follow of execution is prescribed by the dotted arc; (*b*) A *Selector* node with N actions any of which can be selected; (*c*) A *Decorator* node adds additional functionalities to behavior

the subtree could be called or retrieve the status message from the execution of the associated actions. Since the *Decorator* is often transformed into some associated properties of *Action* nodes, we focus on a *canonical* behavior tree that mainly contains the *Sequence*, *Selector* and *Action* nodes. Here we take the popular 2-D fighting game (implemented in the **MUGEN** game engine[1]) for one example of behavior trees.

Fig. 2. A human player is fighting with a *Kung Fu Man* in the **MUGEN** game. The yellow bars above show their health points.

In Fig. 2, the **MUGEN** game hosts two players: one NPC (called *Kung Fu Man*) and one human player. The stage is a 2-D arena where the players can move freely horizontally, and any movement in the vertical axis is achieved by either a jump or a crouching move. The behavior tree as designed in Fig. 3 commands the actions of *Kung Fu Man* in the game-play.

Example 1 (Behavior Tree). *When the game starts, the* Kung Fu Man *chooses either Attack or Defend on executing the Selector node, Choose Mode.*

[1] `http://www.elecbyte.com/`

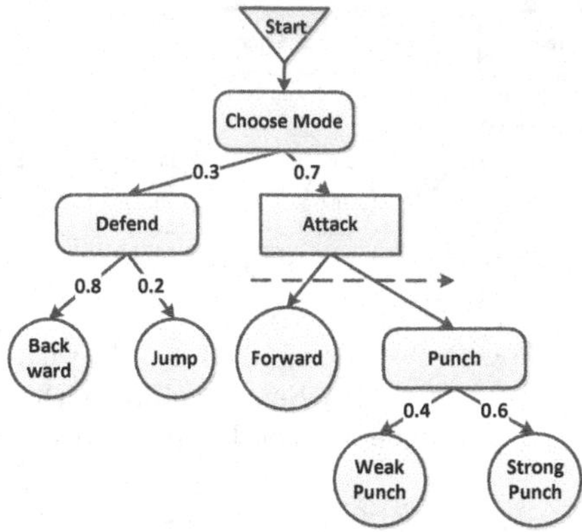

Fig. 3. A behavior tree is designed for an aggressive *Kung Fu Man*

*He is a bit aggressive since there is a large probability (*0.7*) that he attacks a human player. Assume that he selects the Attack mode, he needs to execute a sequence of actions as indicated by the Sequence node, Attack. He will firstly move Forward, and then execute the action Punch if he succeeds in the movement. Finally, he will attack the opponent using one of two punch types. As the probabilities specified in the Selector node, Punch, he may launch a heavy attack with the Strong Punch (that has a larger probability 0.6). The behavior tree returns a success if the* Kung Fu Man *does not fail to execute the actions. Subsequently, a new traversal of the tree will be initiated (at the root node) to control his actions.*

By linking *Selector* and *Sequence* nodes, a behavior tree prescribes a sequence of actions for NPCs in computer games. The configurations of the tree structure (connections of nodes) and parameters (probabilities in the *Selector* node) directly control the NPCs' behavior. In this paper, we mainly exploit the parameter settings for the purpose of designing believable behavior.

3 Believable Behavior Design

Behavior trees provide a simple, scalable and modular solution to design complex NPCs' behavior in computer games. The model settings offer a flexible mechanism to control the behavioral dynamism. By configuring the action probabilities, we expect NPCs to exhibit believable behavior in the game-play. We will firstly formulate the believability of NPCs' behavior as one computational design and analyze its solutions afterwards.

3.1 Computational Believability

A behavior tree structures the relations of action nodes and imposes a set of parameters over actions in a *Selector* node. Formally, we define a behavior tree below.

Definition 1 (Behavior Tree). *Define a behavior tree as:* $\mathcal{T} = \langle \mathcal{V}, \mathcal{E}, \mathcal{P} \rangle$ *where* $\mathcal{V}$ *is a set of Action, Sequence, and Selector nodes;* $\mathcal{E}$ *is a set of edges connecting the nodes; and* $\mathcal{P}$ *is a set of probability distributions guarding the edges in the Selector nodes.*

We assume the known structure of behavior trees including $\mathcal{V}$ and $\mathcal{E}$, and will find a proper setting of $\mathcal{P}$ for designing believable behavior. For the probability distribution, $\mathcal{P}$, we further denote it as $\mathcal{P} = \langle \mathcal{P}_{S_1}, \cdots, \mathcal{P}_{S_M} \rangle$ where $\mathcal{P}_{S_i} = (p_{A_1}, \cdots, p_{A_N})$ is a probability distribution over N actions under the *Selector* node, S_i.

Given a behavior tree $\mathcal{T}$, we may retrieve a set of behavior paths from the tree. Each path is a sequence of actions that an NPC will experience in the game-play. Formally, let $\mathcal{H} = \langle \mathcal{H}_1, \cdots, \mathcal{H}_R \rangle$ be the set of behavior paths where $\mathcal{H}_j = (A_1, \cdots, A_K)$ is a sequence of actions. We define the probability of a path $\mathcal{H}_j$ as:

$$P(\mathcal{H}_j) = \Pi_{k=1}^{K} p(A_k) \tag{1}$$

Here A_k refers to the action labeled either in the *Action* node, or in the *Sequence* node, or in the *Selector* node. Its probability depends on the type of its parent. Consequently, $p(A_k) = 1$ if A_k is a child of either the *Sequence* node or the *Root* node; otherwise, $p(A_k)$ is equal to the probability p_{A_k} as defined for A_k under the corresponding *Selector* node.

Example 2 (Behavior Path). *Given the behavior tree in Fig. 3, we can get 4 behavior paths as:* $\mathcal{H} =<$ $(ChooseMode, Defend, Backward)$, $(ChooseMode, Defend, Jump)$, $(ChooseMode, Attack, Forward, Punch, WeakPunch)$, $(ChooseMode, Attack, Forward, Punch, StrongPunch)$. *The probability of* $\mathcal{H}_3$ *is computed as:* $P(\mathcal{H}_3) = 1 \times 0.7 \times 1 \times 1 \times 0.4 = 0.28$. *The probabilities for all paths are:* $P(\mathcal{H}) = (0.24, 0.06, 0.28, 0.42)$.

Behavior paths, together with the probabilities, specify how an NPC shall act in the game world. The NPC acts intelligently if it is able to learn from its experience. In other words, we can develop intelligent behavior for the NPC by automatically learning its behavior trees. As the set of paths are known given the structure of behavior trees, we need to learn the probabilities, $P(\mathcal{H})$, from game experience. A set of AI/statistical learning techniques have been adapted for this purpose, which is one of the main focuses of AI research in computer games [12]. We may resort to similar techniques that result in the optimal probability values, $P^*(\mathcal{H})$, for intelligent behavior of an NPC.

The probability setting, $P^*(\mathcal{H})$, generates the most intelligent behavior for NPCs since the probabilities are learned from the NPC's experience. The question is: how to measure the intelligence of the NPC's behavior if the NPC executes the behavior with a different probability setting of $P(\mathcal{H})$. Instead of providing a direct measurement, we gauge how its intelligence approaches that of the NPC with the setting of $P^*(\mathcal{H})$. To measure the intelligence gap, we use the distance between two probability distributions, $P^*(\mathcal{H})$ and $P(\mathcal{H})$. Formally, we use the Kullback-Leibler (KL) divergence [8] as defined below.

$$D_{KL}[P(\mathcal{H})||P^*(\mathcal{H})] = \sum_j P(\mathcal{H}_j) ln \frac{P(\mathcal{H}_j)}{P^*(\mathcal{H}_j)} \tag{2}$$

In order to design the intelligent behavior, we need to minimize the distance, D_{KL}, in Eq. 2. The distance converges to zero when the behavior of an NPC achieves the highest intelligence in the optimal setting of $P^*(\mathcal{H})$.

On the other hand, we expect an NPC to execute a broad set of behavior in the game-play. We choose Shannon entropy [5] as the measurement of the behavior diversity. The diversity of the behavior with the probability distribution, $P(\mathcal{H})$, is defined in Eq. 3. A large entropy value indicates more types of actions that an NPC will perform in the real play. Consequently, the NPC's behavior becomes more unpredictable from the eyes of its opponents.

$$E[P(\mathcal{H})] = -\sum_j P(\mathcal{H}_j) ln P(\mathcal{H}_j) \tag{3}$$

As we mentioned, the believability seeks for a good balance of the intelligence and the diversity of NPCs' behavior. The design of believable behavior is to achieve the optimal solutions of the intelligent behavior while to maintain the diversity of the behavior. In other words, we will minimize the KL divergence between $P(\mathcal{H})$ and the optimal one $P^*(\mathcal{H})$, and simultaneously maximize the information entropy of $P(\mathcal{H})$. Formally, we aim to compute the probability distributions, $P(\mathcal{H})$, that are solutions to the optimization problem below.

$$\begin{aligned} \textbf{Objective}: & \ max \\ & BEL = -K_1 \textstyle\sum_j P(\mathcal{H}_j) ln P(\mathcal{H}_j) - K_2 \sum_j P(\mathcal{H}_j) ln \frac{P(\mathcal{H}_j)}{P^*(\mathcal{H}_j)} \\ \textbf{Variables}: & \ P(\mathcal{H}) = \langle P(\mathcal{H}_1), \cdots, P(\mathcal{H}_R) \rangle \\ \textbf{Constraints}: & \ \textstyle\sum_j P(\mathcal{H}_j) = 1 \end{aligned} \tag{4}$$

where $P^*(\mathcal{H})$ are the optimal solutions of intelligent behavior, and K_1 and K_2 are the positive values weighting the diversity and intelligence of behavior respectively in the believability function, BEL.

We compute the solutions, $P(\mathcal{H})$, by applying the partial derivative in the objective function, BEL. Accordingly, the design achieves the optimal believability where an NPC executes the behavior path, $\mathcal{H}_j$, with the probability in Eq. 5.

$$P(\mathcal{H}_j) = \frac{P^*(\mathcal{H}_j)^{\frac{K_2}{K_1+K_2}}}{\sum_j P^*(\mathcal{H}_j)^{\frac{K_2}{K_1+K_2}}} \tag{5}$$

We note that the the probabilities of believable behavior depend on weights between the intelligence and the diversity. An NPC behaves randomly ($P(\mathcal{H}_j) = \frac{1}{R}$) if $K_2 \ll K_1$, and shows the highest intelligence ($P(\mathcal{H}_j) = P^*(\mathcal{H}_j)$) if $K_2 \gg K_1$.

Example 3 (Believability Solutions). *Given a set of 2 behavior paths,* $\mathcal{H} =< \mathcal{H}_1, \mathcal{H}_2 >$, *we learn the path probabilities,* $P^*(\mathcal{H})$=(0.3, 0.7), *for intelligent behavior from game data that record the NPC's performance. We plot the believability function, BEL, for different settings of* K_1 *and* K_2 *in Fig. 4. Selections of* (K_1, K_2) *values balance the factors that contribute into the believability design. Fig. 4(a) favors the intelligence as the dominating attribute of the believability. Hence its solution,* $P(\mathcal{H})$=(0.337, 0.663), *approaches the learned probabilities. On the other hand, Fig. 4(c) attributes the believability to the diversity of actions, and its design,* $P(\mathcal{H})$=(0.479, 0.521), *is close to the random behavior. This may happen to an insane NPC in games. Note that the BEL values are not comparable across different* (K_1, K_2).

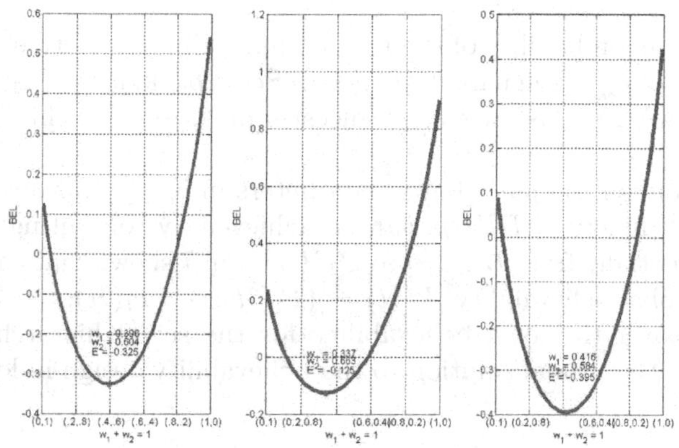

Fig. 4. The probabilities (denoted by red triangles) of believable behavior given K_1 and K_2 values that weight the diversity and the intelligence of behavior respectively

Once we get the path probabilities, $P(\mathcal{H})$, we can compute the probabilities, p_{A_k}, for actions through Eq. 1. This completes the parameter setting of $\mathcal{P}$ for the believable behavior design in behavior trees .

3.2 Bottom-Up Design

As a hierarchical model, a behavior tree is goal-oriented and allows the recursive implementation of complex NPCs' behavior. In practice, it firstly defines a high-level goal (in a *Root* node) which it attempts to achieve, and then links to a set of sub-goals until it constructs primitive actions at the leaves of the tree. Consequently, the tree may become deeply nested and is prone to be asymmetric. Computing the global believability, $P(\mathcal{H})$, for an entire tree involves difficulty in learning the probabilities of intelligent behavior, $P^*(\mathcal{H})$, for a large set of behavior paths.

Following the spirit of modular design in behavior trees, we may compute the local believability, $P_{\mathcal{T}_l}(\mathcal{H})$, for a set of subtrees and incrementally complete the believability design of $\mathcal{T}$ (=$\cup_l \mathcal{T}_l$). To make a further step, we may arrange the subtrees in a hierarchical way and recursively compute the believability in a bottom-up manner.

Let $\mathcal{T}_l$ be level l ($l \in [1, L]$) subtree in $\mathcal{T}$ (the root of $\mathcal{T}$ is in level 0), and $BEL_{\mathcal{T}}$ ($BEL_{\mathcal{T}_l}$) be the global (local) believability function respectively. According to the believability definition, BEL, in Eq. 4, we derive the relations between the global and local believability functions in Eq. 6. It shows that the global believability is the sum of a set of weighted local believability.

$$\begin{aligned} BEL_{\mathcal{T}} = BEL_{\mathcal{T}_1} + P(\mathcal{T}_2)[BEL_{\mathcal{T}_2} + \cdots + \\ P(\mathcal{T}_l)[BEL_{\mathcal{T}_l} + \cdots + P(\mathcal{T}_L)[BEL_{\mathcal{T}_L}] \cdots] \cdots] \end{aligned} \tag{6}$$

where $P(\mathcal{T}_l)$ is the probability of the root node, A_k, in the subtree $\mathcal{T}_l$. The probability is equal to p_{A_k} if the parent of A_k is a *Selector* node in $\mathcal{T}_{l-1}$; otherwise, it is equal to $p_{A'_k}$ where A'_k is the closest ancestor of A_k and is a child of a *Selector* node.

As $P(\mathcal{T}_l)$ is one of the probability parameters in $\mathcal{T}_{l-1}$, solutions to maximize the believability function, $BEL_{\mathcal{T}}$, can be achieved by computing the local believability sequentially from $BEL_{\mathcal{T}_L}$ to $BEL_{\mathcal{T}_1}$. Eq. 7 shows that the optimal solution to the global believability, $P(\mathcal{H})$=<$(P_1(\mathcal{H}), \cdots, P_l(\mathcal{H}), \cdots, P_L(\mathcal{H})$>, can be distributed over the local believability for the set of hierarchical subtrees where $P_l(\mathcal{H})$ is the optimal solution to the believability design in level l subtree, $\mathcal{T}_l$.

$$\begin{aligned} \max_{P(\mathcal{H})} BEL_{\mathcal{T}} = \max_{P_1(\mathcal{H})} [BEL_{\mathcal{T}_1} + P(\mathcal{T}_2) \max_{P_2(\mathcal{H})} [BEL_{\mathcal{T}_2} \\ + \cdots + P(\mathcal{T}_l) \max_{P_l(\mathcal{H})} [BEL_{\mathcal{T}_l} + \cdots \\ + P(\mathcal{T}_L) \max_{P_L(\mathcal{H})} [BEL_{\mathcal{T}_L}] \cdots] \cdots]] \end{aligned} \tag{7}$$

We shall note that the believability design of level l subtree, $\mathcal{T}_l$, needs to find the optimal solutions to maximize the function in Eq. 8. We may show that the analytic solutions, $P_l(\mathcal{H})$, still enjoy the close form similarly in Eq. 5.

$$\max_{P_l(\mathcal{H})} [BEL_{\mathcal{T}_l} + P(\mathcal{T}_{l+1}) OPT(BEL_{\mathcal{T}_{l+1}})] \tag{8}$$

where $OPT(BEL_{\mathcal{T}_{l+1}})$ is the optimal believability value generated from the believability design of $\mathcal{T}_{l+1}$.

In summary, we may decompose the global believability optimization problem into a set of local optimization problems and still achieve the same solutions to the believability in a bottom-up design. The statement is given in Theorem 1.

Theorem 1 (Design Optimality). *Bottom-up design preserves the optimal believability of the global design for an entire behavior tree.*

4 Evaluation and User-Study

We experimented the believability design in the aforementioned **MUGEN** game [2]. We used the *N-Gram* statistical models [18] to learn the probability of intelligent behavior, $P^*(\mathcal{H})$, and computed the probability of believable behavior given different settings of (K_1, K_2) in Eqs. 5 and 1. The resulting probabilities, $p(A_k)$, were used to configure the parameters, $\mathcal{P}$, of the behavior trees. During the game-play, NPCs are controlled by the associated behavior trees. We use NPC(K_1, K_2) to denote the NPC that displays believable behavior given one setting of (K_1, K_2). We report the NPCs' performance when they compete with either other NPCs or human-players. In addition, we invite human players to rank the NPCs in terms of the intelligence and believability of their behavior and advise proper values of (K_1, K_2) in relevant games.

By extending the behavior tree in Fig. 3, we developed five NPCs listed as: NPC$_1(0,1)$, NPC$_2(0.15,0.85)$, NPC$_3(0.25,0.75)$, NPC$_4(0.4,0.6)$, and NPC$_5(0.5,0.5)$. Note that the NPC$_1(0,1)$ is configured with the learned probability, $P^*(\mathcal{H})$, for the believable behavior. In addition, we designed three stereotypes of NPCs (NPC$_A$: *Aggressive*, NPC$_N$:*Neutral*, and NPC$_D$:*Defensive*) that represent typical roles in the **MUGEN** games. The NPCs differ in the probabilities assigned to actions under the *Selector* nodes. We let NPC$_i$ (i=1,$\cdots$,5) start with random actions and compete with the stereotypes individually over 200 matches. During the competition, we had the NPC$_i$ learn from the experience every 20 matches and designed its behavior based on the new probability, $P^*(\mathcal{H})$. We report the total number of matches that the NPC$_i$ won over every stereotype in Table 1.

Table 1. The NPCs (NPC$_1$-NPC$_5$) learn to compete with their opponents (NPC$_A$-NPC$_D$). The more intelligence (NPC$_1$ > $\cdots$ > NPC$_5$) the more matches the NPCs win.

Matches	NPC$_1$	NPC$_2$	NPC$_3$	NPC$_4$	NPC$_5$
NPC$_A$	**181**	176	163	118	72
NPC$_N$	**182**	179	170	127	83
NPC$_D$	**186**	180	170	135	98

Table 1 shows that the NPCs (NPC$_1$-NPC$_5$) perform intelligently when they assimilate most of their learning results given the setting $K_2 > K_1$. They lose few matches if they have fully exploited (where K_2=1) the behavior of their opponents. The results also demonstrate the utility of the *N-Gram* techniques on learning the NPCs' behavior in the game.

We enrolled 27 participants to observe the matches and rank both the intelligence and believability of the NPCs (NPC$_1$-NPC$_5$) when the NPCs were playing

[2] Due to the limited space, we show only the evaluation on the **MUGEN** game while we also conducted study in the popular **StarCraft** Game.

with their opponents. Most of the participants have some experience on the **MUGEN** game. The criteria that they used to evaluate the believability were mainly on plausible sequences of attacks, diverse behavior, and predictable actions. We report the average rankings (with standard deviation) of the NPCs in Table 2. As expected, the $NPC_1(0, 1)$ was ranked as the most intelligent one over all competitions with different types of opponents. However, it lost to $NPC_2(0.15, 0.85)$ on the aspect of the believability.

Table 2. Average rankings of the NPCs (NPC_1-NPC_5) in terms of the intelligence and the believability. 5 is the highest and 1 is the lowest.

NPCs	Criteria	NPC_A	NPC_N	NPC_D
NPC_1	Intelligence	**4.63(0.50)**	**4.38(0.5)**	**4.19(0.66)**
	Believability	3.31(1.14)	3.06(1.44)	2.75(1.34)
NPC_2	Intelligence	4.19(0.75)	4.13(0.96)	4.13(1.02)
	Believability	**4.38(0.88)**	**4.25(0.77)**	**4.06(1.06)**
NPC_3	Intelligence	3.13(0.5)	3.38(0.89)	3.38(1.02)
	Believability	3.44(1.03)	3.88(0.89)	3.94(0.77)
NPC_4	Intelligence	2.06(0.25)	2.13(0.34)	2.31(0.60)
	Believability	2.25(0.77)	2.06(0.25)	2.50(0.97)
NPC_5	Intelligence	1(0)	1(0)	1(0)
	Believability	1.69(1.49)	1.81(1.52)	1.75(1.39)

We made a further step to compare pairs of rankings through *t*-tests. Table 3 shows the *p*-values of the tests between the average rankings of the NPCs. It is a bit surprising that the $NPC_1(0, 1)$ was not perceived as being significantly smarter than the $NPC_2(0.15, 0.85)$. This indicates that the solutions (where K_2=0.85) are sufficient to exhibit the intelligent behavior. Additional diversity of the behavior does not compromise the intelligence, but generate the desired believability in most cases.

Table 3. *p*-values from *t*-tests on the pair comparisons. Entries with an underline are significant at the 95% confidence level.

Criteria	NPCs	NPC_A	NPC_N	NPC_D
Intelligence	$NPC_1 > NPC_2$	0.08	0.23	0.43
	$NPC_2 > NPC_3$	<u>0.00</u>	<u>0.05</u>	0.06
	$NPC_3 > NPC_4$	<u>0.00</u>	<u>0.00</u>	<u>0.00</u>
	$NPC_2 > NPC_4$	<u>0.00</u>	<u>0.00</u>	<u>0.00</u>
Believability	$NPC_2 > NPC_1$	<u>0.00</u>	<u>0.02</u>	<u>0.01</u>
	$NPC_2 > NPC_3$	<u>0.01</u>	0.11	0.37
	$NPC_3 > NPC_4$	<u>0.00</u>	<u>0.00</u>	<u>0.00</u>
	$NPC_2 > NPC_4$	<u>0.00</u>	<u>0.00</u>	<u>0.00</u>

Table 4. Average rankings of the NPCs (NPC_1-NPC_3) in the real-play. 3 is the highest and 1 is the lowest.

Criteria	NPC_1	NPC_2	NPC_3
Intelligence	**2.58(0.51)**	2.25(0.75)	1.17(0.39)
Believability	1.58(0.67)	**2.75(0.62)**	1.67(0.65)

We invited 18 out of 27 participants to play with the top three NPCs (NPC_1-NPC_3) over 100 matches. Subsequently the participants ranked the NPCs according to their personal game experience. In Table 4, the results are consistent with the analysis above. The believability of the NPC_2's behavior is significantly better than that of the others. Most of the participants felt uncomfortable when the $NPC_1(0, 1)$ launched non-breaking attacks with strong punches. It was convincing that the $NPC_2(0.15, 0.85)$ took a light jump after some punches.

5 Conclusion

We propose a computational model for game designers that allow them to create believable behavior for NPCs in computer games. The believability model is rooted in a generic representation of behavior trees and sophisticated AI techniques. We quantify the believability by measuring the intelligence and diversity of NPCs' behavior. In principle, the believability design is seeking for a balance of these two measurements. We further formulate the believability design as one optimization problem and provide analytic solutions to the optimal believability. More importantly, we observe that the bottom-up design can guarantee the optimal believability of the entire behavior tree. This facilitates the practical development on designing believable behavior.

The computational model considers two important attributes (intelligence and diversity of behavior) in the believability design. For future work, we will explore more factors that may contribute into the believability of NPCs' behavior. The challenge is on the development of a quantitative measurement for the attributes. We are more interested in integrating the additional attributes into the established model.

References

1. Amato, C., Shani, G.: High-level reinforcement learning in strategy games. In: Proceedings of the Ninth International Conference on Autonomous Agents and Multiagent (AAMAS), pp. 75–82 (2010)
2. Bates, J.: Virtual reality, art and entertainment. Presence 1(1), 133–138 (1992)
3. Champandard, A.J.: Behavior trees for next-gen game ai. Tutorial, AiGameDev.com (2008)
4. Chang, Y., Maheswaran, R., Levinboim, T., Rajan, V.: Learning and evaluating human-like npc behaviors in dynamic games. In: Proceedings of the Seventh Artificial Intelligence and Interactive Digital Entertainment Conference (AIIDE), pp. 8–13 (2011)

5. Cover, T.M., Thomas, J.A.: Elements of information theory. Wiley-Interscience, New York (1991)
6. Doirado, E., Martinho, C.: I mean it!: detecting user intentions to create believable behaviour for virtual agents in games. In: Proceedings of the Ninth International Conference on Autonomous Agents and Multiagent (AAMAS), pp. 83–90 (2010)
7. Isla, D.: Handling complexity in the halo 2 ai. In: Proceedings of the Fifteenth Conference on Game Developers Conference (2005)
8. Kullback, S., Leibler, R.A.: On information and sufficiency. Ann. Math. Statist. 22(1), 79–86 (1951)
9. Laird, J.E., Newell, A., Rosenbloom, P.S.: Soar: An architecture for general intelligence. Artificial Intelligence 33(1), 1–64 (1987)
10. Langley, P., Choi, D.: A unified cognitive architecture for physical agents. In: Proceedings of the Twenty-First AAAI Conference on Artificial Intelligence (AAAI), pp. 876–881 (2006)
11. Martinho, C., Paiva, A.: Using anticipation to create believable behaviour. In: Proceedings of the Twenty-First National Conference on Artificial Intelligence (AAAI), pp. 175–180 (2006)
12. Rabin, S.: AI Game Programming Wisdom 4. Course Technology (2009)
13. Scott Neal Reilly, W.: Believable Social and Emotional Agents. PhD thesis, School of Computer Science, Carnegie Mellon University (1996)
14. Riedl, M.O., Stern, A.: Believable agents and intelligent scenario direction for social and cultural leadership training. In: Proceedings of the Fifteenth Conference on Behavior Representation in Modeling and Simulation (2006)
15. Russell, S., Norvig, P.: Artificial Intelligence: A Modern Approach, 2nd edn. Prentice-Hall (2003)
16. Tan, C.T., Cheng, H.: Implant: An integrated mdp and pomdp learning agent for adaptive games. In: Proceedings of the Fifth Artificial Intelligence and Interactive Digital Entertainment Conference (AIIDE), pp. 94–99 (2009)
17. Tence, F., Buche, C., De Loor, P., Marc, O.: The challenge of believability in video games: Definitions, agents models and imitation learning. CoRR abs/1009.0451 (2010)
18. Witten, I.H., Bell, T.C.: The zero-frequency problem: estimating the probabilities of novel events in adaptive text compression. IEEE Transactions on Information Theory 37(4), 1085–1094 (1991)
19. Xu, J.Z., Laird, J.E.: Combining learned discrete and continuous action models. In: Proceedings of the Twenty-Fifth AAAI Conference on Artificial Intelligence (AAAI), pp. 1449–1454 (2011)
20. Zeng, Y., Buus, D.P., Hernandez, J.C.: Multiagent based construction for human-like architecture. In: Proceedings of the Sixth International Joint Conference on Autonomous Agents and Multi-Agent Systems (AAMAS 2007), pp. 409–411 (2007)
21. Zeng, Y., Hernandez, J.C., Buus, D.P.: Swarmarchitect: a swarm framework for collaborative construction. In: Proceedings of Genetic and Evolutionary Computation Conference (GECCO 2007), pp. 186–186 (2007)

Discovering Frequent Patterns to Bootstrap Trust

Murat Sensoy[1,2], Burcu Yilmaz[1,3], and Timothy J. Norman[1]

[1] Department of Computing Science, University of Aberdeen, UK
[2] Ozyegin University, Istanbul, Turkey
[3] Gebze Institute of Technology, Kocaeli, Turkey
{m.sensoy,burcu,t.j.norman}@abdn.ac.uk

Abstract. When a new agent enters to an open multiagent system, bootstrapping its trust becomes a challenge because of the lack of any direct or reputational evidence. To get around this problem, existing approaches assume the same a priori trust for all newcomers. However, assuming the same a priori trust for all agents may lead to other problems like *whitewashing*. In this paper, we leverage graph mining and knowledge representation to estimate a priori trust for agents. For this purpose, our approach first discovers significant patterns that may be used to characterise trustworthy and untrustworthy agents. Then, these patterns are used as features to train a regression model to estimate trustworthiness. Lastly, a priori trust for newcomers are estimated using the discovered features based on the trained model. Through extensive simulations, we have showed that the proposed approach significantly outperforms existing approaches.

1 Introduction

In open systems like the Web, there is no central authority that monitors interacting agents and guarantees that every agent in the system behave as expected. For instance, a seller in an e-market place such as *ebay* may list a product at cheaper price but may not deliver the same product or any thing at all. This brings the necessity of agents to evaluate others and select the most trustworthy interaction partners among alternatives. While the word trust may have different definitions in different domains, we define it here pragmatically as the degree of belief or subjective probability, with which a trustor believes a trustee will perform as expected when relied upon [12].

Existing approaches for trust generally depend on an interaction history to compute trust of an agent x to another agent y. The interaction history contains the interactions between x and y (i.e., direct evidence) and their outcomes (e.g., *success* and *failure*). If the number of the direct interactions is not enough to compute trust with confidence, reputational evidence is used to compute the trustworthiness of y. The past interactions between other agents and y serve as reputational evidence. Based on direct and reputational evidence, statistical trust approaches like Beta Reputation System (BRS) [11] and $TRAVOS$ [17] is used to compute the trustworthiness of the agent y as the subjective probability that a future interaction with y would have desirable outcome. As the number of evidence increases, the accuracy of the estimated trustworthiness increases and the number of unsatisfactory interactions are minimised.

Although the existing approaches can accurately model the trust, they require repetitive interactions between the agents to build an interaction history. This requirement leads to two related problems: *bootstrapping* and *whitewashing*. The bootstrapping

L. Cao et al.: ADMI 2012, LNAI 7607, pp. 93–104, 2013.

problem arises when a new agent joins to the system. In this case, trust cannot be computed based on direct or reputational evidence. To deal with bootstrapping problem, the existing approaches assume a priori trust value for the newcomers. For instance, in BRS and $TRAVOS$, the a priori trust is 0.5, which means positive and negative outcome is equally likely. If the a priori trusts is high, the agents with bad reputations whitewash their bad reputation and enters to the society as a newcomer. On the other hand, if the a priori trust is too low, the newcomers may not have any chance to interact with others and build a good reputation.

Malicious agents may adopt certain behavioural patterns to achieve their goals. These patters may determine their choices and lead to the emergence of certain motifs in their relationships with other entities. Even if an agent whitewashes its identity and change/forge some of its observable attributes, the same motifs may be observed as long as the agent does not change its behaviour. For example, a malicious seller may change its identity and advertise a completely new profile whenever its reputation decreases. Although the *name*, *location*, *email*, and *web site* in the new profile are different from previous ones, all these email addresses and web sites could be hosted by the same or similar service providers (e.g., free hosting services). If malicious sellers have a tendency to bear the same or similar pattern in their profiles, we may build a stereotype such as "sellers using free hosting services are less trustworthy". This stereotype is not based on a simple attribute of seller (e.g., a specific email address), but an example of a complex feature discovered about the malicious sellers.

This paper proposes to discover and exploit the complex features (i.e., patterns) of agents [3] to estimate a priori trust for them. Knowledge about each known agent is represented in detail as a graph based on an ontology. Then, these graphs are mined to discover two groups of patterns that frequently appear in trustworthy and untrustworthy agents respectively. Lastly, the discovered patterns are used as features to train a regression model to estimate the a priori trust for the unknown agents. Through extensive simulations, we have showed that the proposed approach significantly outperforms existing trust approaches.

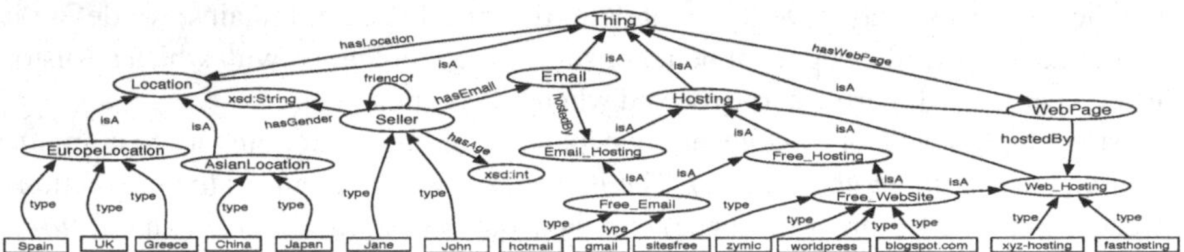

Fig. 1. A part of a simple ontology to describe sellers, their properties and relationships

2 Modelling Trust

Several approaches have been proposed to model trust in the literature [12]. A number of these approaches are based on *Subjective Logic* (SL), which is a belief calculus that allows agents to express opinions as degrees of belief, disbelief and uncertainty about propositions. Let ρ be a proposition such as "information source y is trustworthy in context c". Then, the binary opinion of agent x about ρ is equivalent to a Beta distribution. That is, the binomial opinion about the truth of a proposition ρ is represented as the tuple (b, d, u, a), where b is the belief that ρ is true, d is the belief that ρ is false, u is the uncertainty, and a is base rate (a priori probability in the absence of evidence), as well as $b + d + u = 1.0$ and $b, d, u, a \in [0, 1]$. Opinions are formed on the basis of positive and negative evidences, possibly aggregated from different sources. Let r and s be the number of positive and negative past observations about y respectively, regarding ρ. Then, b, d, and u are computed based on Equation 1.

$$b = \frac{r}{r+s+2}, \; d = \frac{s}{r+s+2}, \; u = \frac{2}{r+s+2} \tag{1}$$

Then the opinion's probability expectation value is computed using Equation 2. Considering ρ, the computed expectation value can be used by x as the trustworthiness of y in the context c [12].

$$t^{x}_{y:c}(r, s, a) = b + a \times u = \frac{r + a \times 2}{r + s + 2} \tag{2}$$

The base rate parameter a represents a priori degree of trust x has about y in context c, before any evidence has been received. The default value of a is mostly choose as 0.5 in literature [10], which means that before any positive or negative evidence has been received, both outcomes are equally likely. While x has more evidence to evaluate trustworthiness of y, the uncertainty u, so the effect of a, decreases.

For clarity, in this paper, we assume that the trust is computed by the same trustor agent x in the same context c. Therefore, we shortly use t_y instead of $t^{x}_{y:c}$ to represent trustworthiness of y in the context c for the agent x.

3 Knowledge Representation

To describe agents and their relationships semantically and flexibly, we propose to use an ontology [9]. To demonstrate toy examples in the paper, Figure 1 shows a part of a simple ontology, where arcs, ellipses, and rectangles represent relationships, concepts and their instances, respectively. Description of an agent is represented using $\langle subject$, $relation$, $object \rangle$ triples, such as $\langle john$, $hasLocation$, $Spain \rangle$. In these triples, subjects, objects, and properties are terms from the ontology. That is, subjects are instances of concepts (e.g., *john*); objects are literals (e.g., 60), concepts (e.g., *Location*), or their instances (e.g., *Spain*); and relations are dataype properties (e.g. *hasAge*) or object properties (e.g., *hasLocation*) [9].

Each known agent is represented as an instance (e.g., instance of *Seller* concept) in the ontology and described using $\langle subject$, $relation$, $object \rangle$ triples. The description of the agent y can be semantically represented as a labelled directed graph G_y. The node representing y in G_y is called terminal node and connected to other nodes through relationships from ontology. Nodes in G_y refers to either literals or instances and edges of G_y correspond to relationships between those. A seller agent $john$ is represented as the graph shown in Figure 2. Given the ontology, this graph describes the seller $john$. However, it is not possible to completely interpret or reason about it without the ontology, since the terms (i.e., individuals and properties) used in this graph are defined within this ontology. That is why this graph is called *ontology-dependent* graph.

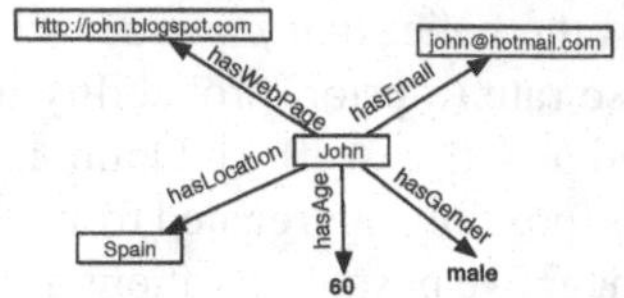

Fig. 2. Ontology-dependent description of the seller $john$

All most all of state of the art graph mining tools do not incorporate ontological knowledge [8]. Hence, they completely neglect the semantics of the labels in a graph. For instance, for these approaches, $Spain$ in the graph of Figure 2 is just a label of a node. However, the ontology of Figure 1 implies that $Spain$ is a location in Europe. Therefore, ontology-dependent graphs are not much informative for existing graph mining approaches [18]. One way of making them more informative as stand alone graphs is to embed relevant ontological knowledge into them. For this purpose, we use an ontological reasoner such as Pellet [16] and derive all direct and indirect statements (i.e., triples) about the individuals in the graph. Let us note that these individuals correspond to nodes in the original graph. Then, based on each triple $\langle subject, relation, object\rangle$, we create new nodes and relations if they are missing in the graph. As a result, the graph contains all direct and inferred information about the case at different levels of abstraction. Using individual's names as nodes' labels may hamper frequent pattern mining, since these individuals may not appear frequently in the dataset. To avoid this, we replaced these labels with "?" and added name (i.e., URI) of the referred individuals as a property to these nodes, i.e., using *name* datatype property. The resulting graph is called *ontology-independent* graph and referred to as g_y, because we do not need an ontology to infer properties of individuals on the graph; instead each node referring to an individual bares all direct and inferred attributes of the individual. Figure 3 shows ontology-independent version of the case graph shown in Figure 2. The terminal node is coloured *black* in the graph.

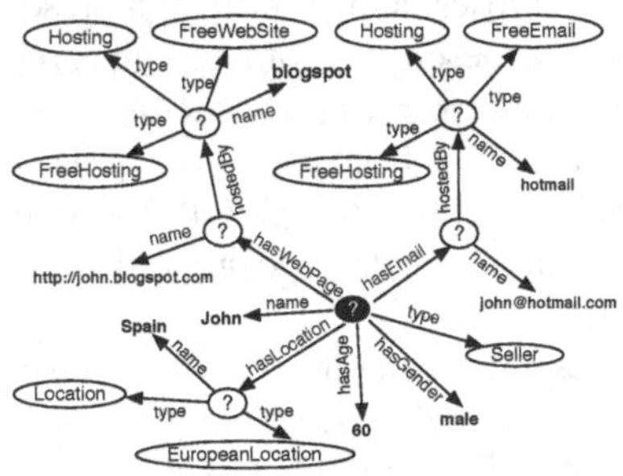

Fig. 3. Ontology-independent description of the seller $john$

Although we use a simple toy example to demonstrate the proposed graph-based representation here, this representation is very expressive and flexible enough to accommodate not only the attributes of the trustee but also all of its relationships with other entities (e.g., *friend of friend* relationships).

4 Graph Mining for Estimating a Priori Trust

In the previous section, we have described how the information about an agent can be represented as a graph, which can capture various aspects of the agent such as its attributes and even its relationships with other agents at different levels of abstraction. In this section, we propose to exploit graph mining techniques to discover frequent patterns that exist more frequently in either trustworthy or untrustworthy agents. For this purpose, we create graph datasets composed of graphs describing agents. Then, we use the discovered frequent patterns as features to estimate a priori trust for the agents.

4.1 Determination of Significant Patterns

For each known agent y, a graph g_y is generated to describe the agent. Then, g_y is labelled based on t_y; the labels are the categories based on the degree of trustworthiness. In this work, for the sake clarity and simplicity, we use only two categories: *trustworthy* and *untrustworthy*. Therefore, in the resulting graph data set, there are two classes of graphs: C_+ and C_-, which represent the graphs generated from *trustworthy* and *untrustworthy* agents respectively. An agent y is considered trustworthy if $t_y \geq \gamma$ and untrustworthy if $t_y \leq \delta$ where $\gamma > 0.5 > \delta$[1]. We want to discover two sets of patterns: P_+ and P_-, which are called patterns of *trust* and *distrust*, respectively. A pattern $p \in P_-$ is a subgraph that repeats frequently in the graphs of C_- and rarely in the graphs of C_+. Therefore, p distinguishes untrustworthy agents from trustworthy ones. Similarly, P_+ represents the significant patterns that repeat more frequently in the graphs of C_+. A formal definition of the patterns in P_- and P_+ is given in Definition 1 based on the frequencies explained in Definition 2.

Definition 1. Let $\nu \in \{+,-\}$ and $\bar{\nu} \in \{+,-\} \setminus \{\nu\}$. A frequent pattern $p \in P_\nu$ is a pattern such that $f(p, C_\nu) \geq \alpha_\nu > \sigma f(p, C_{\bar{\nu}})$, where $f(p, C_\nu)$ and $f(p, C_{\bar{\nu}})$ are frequencies of p in classes C_ν and $C_{\bar{\nu}}$, respectively; $0 < \alpha_\nu \leq 1$ is a threshold and $\sigma \geq 1$ is a coefficient. ∎

Definition 2. Given a pattern p, $f(p, C_\nu)$ is the frequency of the pattern in the graphs of class C_ν and computed by the formulae:

$$f(p, C_\nu) = \frac{N_{p:C_\nu}}{|C_\nu|}$$

where $N_{p:C_\nu}$ and $|C_\nu|$ are the number of graphs containing p in class C_ν and size of C_ν, respectively. ∎

Without support constraints (i.e., α_+ and α_-), we have to deal with huge number of patterns. To narrow down the search space, the trustor agent determines these constraints. An additional constraint can be the *minimum size* of the frequent patterns, where the size of the pattern is determined by its number of edges. In the literature, a number of *frequent subgraph mining* approaches have been proposed to extract frequent patterns in multi-class graph datasets using support and size constraints [5, 6, 18].

[1] To accommodate the blurred boundary between of trustworthiness and untrustworthiness, we set γ and δ to 0.75 and 0.4, respectively, in our evaluations.

4.2 Estimating a Priori Trust

As described in detail above, the trustor agent finds two sets of frequent patterns (P_+ and P_-) using frequent subgraph mining. In this section, we describe how these significant patterns can be used as features to estimate a priori trust for agents. Then, the estimated a priori trust is used as the base rate a in Equation 2 while computing the trustworthiness of the agents.

To estimate the a priori trust, the trustor can use various machine learning techniques [1]. In this work, we employ M5 regression tree algorithm [15] for this purpose. Given a training set and a set of features $\mathcal{F}$, this algorithm learns a mapping between feature values of an agent and its trustworthiness. In this paper, each feature $f_j \in \mathcal{F}$ corresponds to a specific pattern $p_j \in P_+ \cup P_-$ and its value for an agent i indicates if the graph g_i describing the agent entails p_j or not.

The training set is prepared using the descriptions of known agents and their trustworthiness. That is, for each known agent i, the training set contains a feature vector $\boldsymbol{v_i}$ and trust value t_i. Each field v_{ij} in the feature vector corresponds to the value of feature f_j for the agent i, which is determined by checking entailment of p_j by g_i. That is, $v_{ij} = 0$ if the patter p_j does not exist in the graph g_i describing i; otherwise $v_{ij} = 1$. In this way, we transform agents' graphs of various sizes into a vector dataset with fixed dimensions. After creating feature vectors for each known agent i, the set of $(\boldsymbol{v_i}, t_i)$ pairs are used as a training set to learn the regression model. The trained regression model is used as a function $R : \boldsymbol{v} \rightarrow [0, 1]$ that takes the feature vector $\boldsymbol{v}$ of an agent as input and returns an estimation of its trustworthiness. This estimation is not based on any evidence about the agent but only its features, hence what is estimated is actually the agent's a priori trust.

In order to compute trustworthiness of an unknown agent y, first the a feature vector $\boldsymbol{v_y}$ is created based on the discovered features. Then, t_y is computed using Equation 2 where $R(\boldsymbol{v_y})$ is used as the base rate a.

The trustor may have more interactions over time and learn trustworthiness of others better based on new evidence. Meanwhile, dynamics of the society may change and new behavioural patterns may be adopted by malicious agents. Therefore, the trustor may periodically use the described approach to discover new features significant for trustworthiness and retrain the regression model based on these new features.

5 Evaluation

In evaluating our approach, we employed a simulated agent society where a set of trustor agents interact with a set of trustee agents over a number of rounds. Each trustee is assigned a performance profile which determines how it will behave. Each profile specifies the mean and standard deviation parameters of a Gaussian distribution from which simulated interaction outcomes will be drawn. A trustor considers an interaction's outcome as a *success* if it is greater than a threshold λ; otherwise it is considered a *failure*. This threshold could vary for each trustor, so that different trustors may perceive the same outcome differently. However, for simplicity, we assume that all trustors use the same threshold value. Table 1 uses the values of all parameters used in the simulations.

Table 1. Experimental Parameters

Parameter	Value	Description
γ	0.75	Threshold for trustworthiness
δ	0.4	Threshold for untrustworthiness
$\alpha_- = \alpha_+$	0.1	Support constraints
σ	2	Frequency coefficient
$Life$	500	Simulation life
N_a	100	Number of trustee agents
N_t	20	Number of trustor agents
N_i	100	Total number of concept instances
N_{avail}	10	Number of available trustees
N_{rp}	10	Number of reputation providers
Δ	10	Learning interval
λ	0.5	Success threshold
ψ	[0, 1]	Probability of leaving society

In our experiments, we have associated each profile with certain attributes and patterns. Using the proposed approach, the trustor tries to learn this association and exploits it to bootstrap trust. Table 2 lists the profiles used in our evaluations.

Table 2. Profiles and performance properties

Profile ID	Mean	SDV
P_1	0.9	0.05
P_2	0.6	0.15
P_3	0.4	0.15
P_4	0.3	0.1
P_5	0.0	1

Each trustee y has a set of attributes, each defined as a triple $\langle y, hasRel, i_{k:j}\rangle$ (shortly $hasRel(y, i_{k:j})$ hereafter) where $hasRel$ is the object property used to define attributes and $i_{k:j}$ is the jth instance of a leaf concept C_k in the concept hierarchy. The hierarchy of concepts used in our simulations are shown in Figure 4.

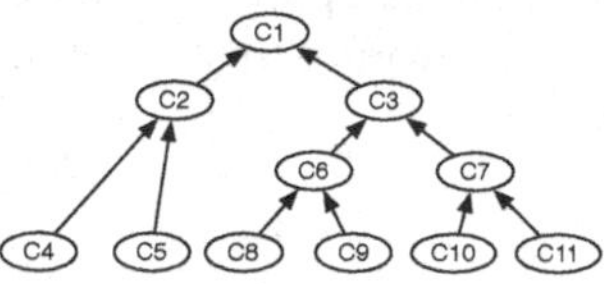

Fig. 4. Concept hierarchy used in simulations

We have 20 trustors and 100 agents (i.e., trustees) in our simulations. Each profile has equal number of trustees and each trustee can belong to one profile. Each simulation is run for 500 discrete time steps. At each step, each trustor selects one trustee to interact with. For this purpose, it gets the list of 10 randomly selected available trustees in the environment, evaluate their trustworthiness using a trust model and selects one with the highest trustworthiness. We have compared the following trust models in our evaluations.

- **Beta Reputation System** (BRS)**:** Trust is estimated using Equation 2 where a is set to 0.5 [11].
- **Stereotypical Trust** (ST)**:** Trust is estimated using Equation 2 where a is computed using a M5 regression model, which is trained using observable attributes of known agents and their trustworthiness [2].
- **Trust through Pattern Discovery** (TPD)**:** Trust is estimated using the proposed approach.

The trustor also obtains a list of 10 reputation providers from the environment and queries them for evidence (i.e., past interaction outcomes) about trustees. In this work, other trustors serve as reputation providers and we assume they honestly share their evidence about trustees. This assumption is made only to focus on bootstrapping trust in this work. Handling deceptive evidence is out of the scope of this paper, but our approach be extended to handle deceptive evidence as described in [7].

To simulate the dynamical of the environment, we introduce the parameter ψ that determines the probability that a trustee leave the society. When a trustess leaves the society, a new trustee of the same profile joins the society. In this way, we maintain the balance of profiles and also simulate *whitewashing* behaviour of trustees.

In the following sections, we evaluate our approach against three settings: i) performance of trustees are not correlated with their descriptions, ii) their performances are correlated with their attributes, and iii) their performances are correlated with the patterns in their descriptions. We repeated our experiments 5 times and report their average results. As performance metric, we use the ratio of successful interactions. For frequent subgraph mining, we have used ParMol [14].

5.1 No Discriminative Features

In this setting, each trustee is assigned randomly a set of attributes. Hence, there is no correlation between participants' trustworthiness and their features. Figure 5 shows our results for this setting when no trustee leaves the society through out experiments ($\psi = 0.0$). In this setting, all of the three approaches have the same performance; they successfully determine the most trustworthy trustees in the environment and lead high ratio of successful interactions.

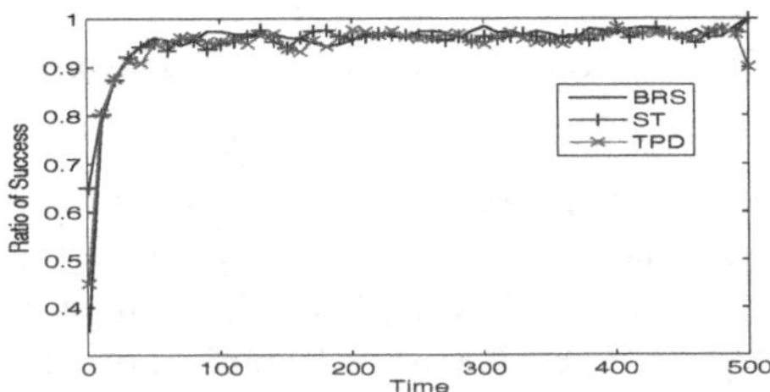

Fig. 5. Ratio of success when there is no discriminative features ($\psi = 0.0$)

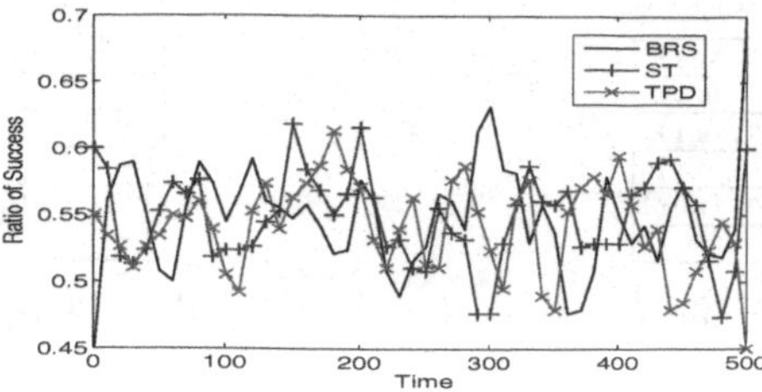

Fig. 6. Ratio of success when there is no discriminative features ($\psi = 0.5$)

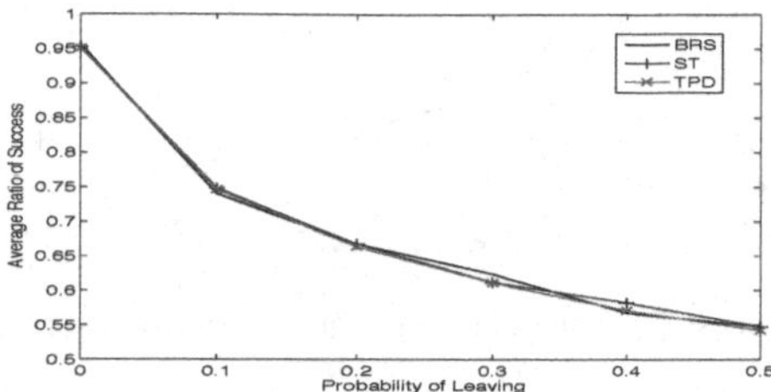

Fig. 7. Ratio of success as ψ changes when there is no discriminative features

Figure 6 shows our results for this setting when ψ is increased to 0.5. In this setting, trust approaches do not have enough interaction history to learn trustworthiness of each trustee. Hence, they have much lower performance compared to the case $\psi = 0$.

We repeated our experiments for different ψ values. Our results are shown in Figure 7. Our experiments indicate that the proposed approaches does not do worse than the existing approaches when there is no correlation between features of trustees and their performance. In such settings, the statistical trust approaches fail mostly because of the lack of enough evidence about the trustworthiness.

5.2 Attributes as Discriminative Features

In this setting, we associate the attributes in Table 8 to profiles as shown in Table 9. In this way, we set all trustees sharing the same profile have some common attributes in addition to their randomly assigned attributes.

Figure 10 shows our results for this setting for different values of ψ. The figure indicates that the performance of BRS significantly decreases as the probability of leaving the society increases. BRS assumes the same a priori trust for all trustees and build trustworthiness for individual trustees over time. However, as ψ increases, BRS could not have enough evidence to compute trust precisely. Unlike BRS, ST learns stereotypes about trustworthiness based on the attributes of trustees. Hence, it exploits the correlation between attributes of trustees and their attributes to precisely estimate a priori trust for trustees. As a result, ST have a very high success ratio that does not decrease with ψ. TPD successfully discovers the patterns implied by the attributes in Table 8 and uses these patterns as features to learn a mapping between features and a priori trust. As a result, in this setting, TPD is as successful as ST and correctly estimates trustworthiness even at high values of ψ.

a_1	$hasRel(?x, i_{4:1})$
a_2	$hasRel(?x, i_{5:1})$
a_3	$hasRel(?x, i_{8:1})$
a_4	$hasRel(?x, i_{9:1})$
a_5	$hasRel(?x, i_{10:1})$
a_6	$hasRel(?x, i_{11:1})$

Fig. 8. Attribute definitions

Profile ID	a_1	a_2	a_3	a_4	a_5	a_6
P_1	X					X
P_2		X		X		
P_3			X	X		
P_4		X	X		X	
P_5		X	X			X

Fig. 9. Profiles descriptions

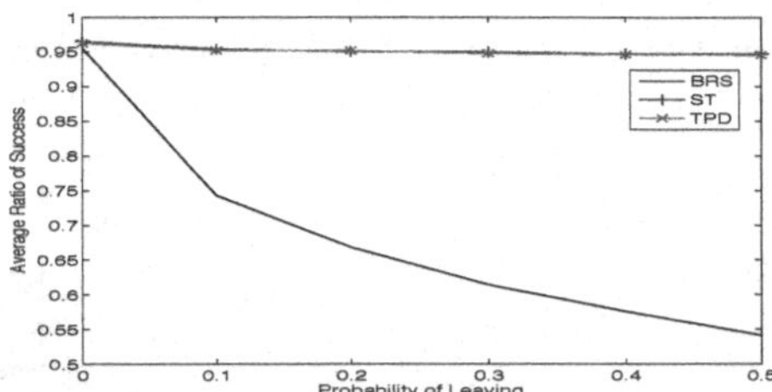

Fig. 10. Ratio of success as ψ changes when attributes are discriminative features

5.3 Patterns as Discriminative Features

In this setting, we associate the pattern in Table 3 to profiles as shown in Table 4. In this way, we set all trustees sharing the same profile have some common patterns.

In this challenging setting, trust approaches like BRS and ST could not estimate trustworthiness correctly as ψ increases. ST mainly fails because it could not find any mapping between attributes of trustees and their trustworthiness. On the other hand, the trustees belonging to the same profile share patterns. TPD discovers these patterns correctly and uses them to learn the correlation between these features and trustworthiness of trustees. Our results clearly show that TPD can estimate trustworthiness very successfully even in highly dynamic environments.

Table 3. Patterns and their definitions

p_0	$hasRel(?x, ?a) \wedge type(?a, C_1)$
p_1	$hasRel(?x, ?a) \wedge type(?a, C_2) \wedge hasRel(?x, ?b) \wedge type(?b, C_3)$
p_2	$hasRel(?x, ?a) \wedge type(?a, C_4) \wedge hasRel(?x, ?b) \wedge type(?b, C_6)$
p_3	$hasRel(?x, ?a) \wedge type(?a, C_5) \wedge hasRel(?x, ?b) \wedge type(?b, C_7)$
p_4	$hasRel(?x, ?a) \wedge hasRel(?a, ?b) \wedge hasRel(?b, ?c) \wedge type(?c, C_3)$
p_5	$hasRel(?x, ?a) \wedge hasRel(?a, ?b) \wedge hasRel(?b, ?c) \wedge type(?c, C_6)$
p_6	$hasRel(?x, ?a) \wedge hasRel(?a, ?b) \wedge hasRel(?b, ?c) \wedge type(?c, C_7)$

Table 4. Profiles described using patterns of Table 3

Profile ID	p_1	p_2	p_3	p_4	p_5	p_6
P_1	X	X				
P_2	X		X			
P_3				X		
P_4				X	X	
P_5				X		X

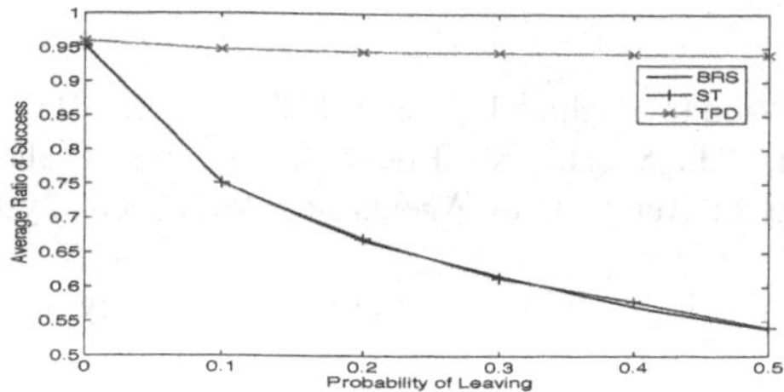

Fig. 11. Ratio of success as ψ changes when patterns are discriminative features

6 Discussion

There are a couple of statistical models for computing trust and reputation in multiagent systems. The beta reputation system (BRS) is proposed by Jøsang and Ismail [11]. It estimates reputations of service providers using Subjective Logic, where the trust is modelled using the beta probability density function. TRAVOS is proposed by Teacy *et al.* [17]. Similar to BRS, it uses beta probability density functions to compute consumers' trust on service providers. Caverlee *et al.* [4] propose the *SocialTrust* framework for tamper-resilient trust establishment in online social networks. In this framework, initially all users have the same level of trust. Then, SocialTrust dynamically revise trust ratings based on the interaction history.

The approaches mentioned above use direct or indirect past interactions with the other agents to compute trust. However when a new agent enters to a society, there is no direct or reputational evidence about the agent. Hence, bootstrapping trust becomes a challenge in open and dynamic multiagent systems. To address this challenge, Liu *et al.* propose agents to form stereotypes using their previous transactions with others [13]. In their approach, a stereotype contains certain observable attributes of agents and an expected outcome of the transaction. Similarly, Burnett *et al.* proposed to use stereotyping to bootstrap trust evaluations based on Subjective Logic [2]. Their approach allows agents to generalise their experience with known agents as stereotypes and apply these when evaluating new and unknown agents. Stereotypes are learned with standard *M5 regression tree* algorithm using a training set composed of observable attributes of known agents and their trustworthiness.

In this paper, we argue that agents with similar behaviour may share some patterns in their descriptions or relationships. We propose to discover these significant patterns in trustworthy and untrustworthy agents and exploit them to learn bootstrapping trust in dynamic and uncertain environments. Through extensive simulations, we show that the proposed approach is at least as good as the existing approaches in all settings. However, it significantly outperforms existing approaches when agents with similar behaviour share some common patterns. As a future work, we would like to use our approach to estimate trust in social networks with real data.

References

1. Alpaydin, E.: Introduction to Machine Learning. MIT Press (2001)
2. Burnett, C., Norman, T.J., Sycara, K.: Bootstrapping trust evaluations through stereotypes. In: Proceedings of Autonomous Agents and Multiagent Systems (AAMAS 2010), pp. 241–248 (2010)
3. Cao, L., Gorodetsky, V., Mitkas, P.: Agent mining: The synergy of agents and data mining. IEEE Intelligent Systems 24(3), 64–72 (2009)
4. Caverlee, J., Liu, L., Webb, S.: Towards robust trust establishment in web-based social networks with SocialTrust. In: WWW 2008: Proceeding of the 17th International Conference on World Wide Web, pp. 1163–1164 (2008)
5. Chakrabarti, D., Faloutsos, C.: Graph mining: Laws, generators, and algorithms. ACM Comput. Surv. 38 (June 2006)
6. Cook, D.J., Holder, L.B.: Graph-based data mining. IEEE Intelligent Systems 15(2), 32–41 (2000)
7. Şensoy, M., Zhang, J., Yolum, P., Cohen, R.: Poyraz: Context-aware service selection under deception. Computational Intelligence 25(4), 335–366 (2009)
8. Donato, D., Gionis, A.: A survey of graph mining for web applications. In: Aggarwal, C.C., Wang, H., Elmagarmid, A.K. (eds.) Managing and Mining Graph Data, pp. 455–485 (2010)
9. W.O.W. Group. OWL 2 web ontology language: Document overview (2009), http://www.w3.org/TR/owl2-overview
10. Jøsang, A.: Subjective Logic. Book Draft (2011)
11. Jøsang, A., Ismail, R.: The beta reputation system. In: Proceedings of the Fifteenth Bled Electronic Commerce Conference e-Reality: Constructing the e-Economy, pp. 48–64 (June 2002)
12. Jøsang, A., Ismail, R., Boyd, C.: A survey of trust and reputation systems for online service provision. Decis. Support Syst. 43, 618–644 (2007)
13. Liu, X., Datta, A., Rzadca, K., Lim, E.-P.: Stereotrust: a group based personalized trust model. In: Proceeding of the ACM Conference on Information and Knowledge Management, pp. 7–16 (2009)
14. Meinl, T., Wörlein, M., Urzova, O., Fischer, I., Philippsen, M.: The ParMol package for frequent subgraph mining. In: Electronic Communications of the EASST, pp. 94–105 (2007)
15. Quinlan, J.: Learning with continuous classes. In: Proceedings of the 5th Australian Joint Conference on Artificial Intelligence, Singapore, pp. 343–348 (1992)
16. Sirin, E., Parsia, B., Grau, B.C., Kalyanpur, A., Katz, Y.: Pellet: A practical OWL-DL reasoner. Web Semant. 5(2), 51–53 (2007)
17. Teacy, W., Patel, J., Jennings, N., Luck, M.: TRAVOS: Trust and reputation in the context of inaccurate information sources. Autonomous Agents and Multi-Agent Systems 12(2), 183–198 (2006)
18. Yan, X., Han, J.: gSpan: Graph-based substructure pattern mining. In: Proceedings of the International Conference on Data Mining, pp. 721–724 (2002)

Subjectivity and Objectivity of Trust

Xiangrong Tong[1], Wei Zhang[1], Yu Long[1], and Houkuan Huang[2]

[1] School of Computer Science, Yantai University, Shandong 264005, China
{txr,zw}@ytu.edu.cn

[2] School of Computer and Information Technology, Beijing Jiaotong University
Beijing 100044, China
hkhuang@bjtu.edu.cn

Abstract. Trust plays an important role in the fields of Distributed Artificial Intelligence (DAI) and Multi-agent Systems (MAS), which provides a more effective way to reduce complexity in condition of increasing social complexity. Although a number of computational issues about trust have been studied, there has to date been little attempt to investigate the differences between Objective Trust (OT) and Subjective Trust (ST). In this paper, we will rectify this omission. Particularly, we study the relationship between OT and ST, and propose Transitive Trust (TT) based on ST. We show that, differing with OT, ST is related to preferences of agents. We propose three rules to form trust framework, and give an example to illustrate the process of trust formation. We finally characterize some useful properties of OT and ST.

Keywords: Multi-agent System, Subjective Trust, Objective Trust, Transitive Trust, Recommendation.

1 Introduction

Nowadays, both Agent Technology and Data Mining technologies have reached an acceptable level of maturity. A fruitful synergy of the two technologies has already been proposed, that would combine the benefits of both worlds and would offer computer scientists with new tools in their effort to build more sophisticated software systems [1,2,3]. In conditions of increasing social complexity, an agent can make much efforts by data mining to overcome the complexity of choices. Particularly, trust constructs a more effective form than utility theory of complexity reduction [4].

As Teacy and Khosravifar [5,6] defined, an agent's (trustor) trust in another (trustee) is defined as the measure of willingness that trustee will fulfill what he agrees to do and computed by considering personal interaction experiences and collecting suggested ratings from others. We believe that trust has three modes which are Objective Trust (OT), Subjective Trust (ST) and Transitive Trust (TT). We assume that trust only based on personal interaction experiences is OT, and ST means that an agent put some personal preferences on interactions, while trust relation formed without interactions is TT which is recommended from other agents.

L. Cao et al.: ADMI 2012, LNAI 7607, pp. 105–114, 2013.

Most of existed researches are based on probability computational models of trust. It is not consistent with the definition of trust as Teacy and Khosravifar announced, as well as lack of strong persuasion. Particularly, ST is a common issue. Because ST is not a transitive relationship due to the subjectivity of trust, trust relay is not a reliable way in our society. Actually, inaccurate information provided by others is more or less due to the subjectivity of trust.

Unfortunately, little efforts were done for considering ST in previous work. Furthermore, no attempts has been done to bridge the gap of OT and ST. A trust network can not be built up because of non-transitivity of ST. We hope to investigate the subjectivity and objectivity of trust thoroughly. Some important properties of trust are proposed in the following sections. We also consider TT, the transitivity of ST, in our society to form trust network.

The first contribution of this paper is the proposition of subjectivity and objectivity of trust. Therefore, preferences and the transitivity of ST are considered systemically. The second contribution is that some useful properties are also given. It may help the future researches in some aspects.

2 Basic Notions

We assume that under a cooperative environment several agents, which have individual interaction states, form a society Ag. Some of agents interacted with each other inside this society. Therefore, based on objective interaction results and subjective preferences, trust relation will be formed.

We Assume that the environment where agents interacted is stable, and we assume that we discuss only a fixed attribute of agents in this paper and the attribute is invariable, i.e. all of agents interact in a fixed environment and on the same issue all the while.

First of all, let OT and ST translate from multiple value to binary value in this paper to discuss the following properties. Therefore, for $\forall a_i, a_j \in Ag$, either $\langle a_i, a_j \rangle \in OT$ (it means that a_i trusts a_j)or $\langle a_i, a_j \rangle \notin OT$ (it means that a_i does not trust a_j) is true. i.e. for $\forall a_i, a_j \in Ag$, $OT_i(j) = \{0,1\} \wedge ST_i(j) = \{0,1\}$.

In order to take advantage of the data of agent's interactions, we give informational states of individual agents [12].

Definition 1. Informational states of individual agents

$$V_i(j) = \langle v_i^1(j), v_i^2(j), \ldots, v_i^{s_i^j}(j) \rangle$$

where, $V_i(j)$ is a vector which saves interaction history between agent a_i and agent a_j ranked by time in a fixed situation.

s_i^j denotes the quantity of interactions between agent a_i and agent a_j in a fixed situation. Clearly $s_j^i = s_i^j$.

$v_i^l(j)$ denotes the result of lth interaction between agent a_i and a_j in a fixed situation. $l = 1, 2, \ldots, s_i^j$. Under a cooperative environment, the result of interaction is same to the bilateral agents, i.e. if the interaction is successful, then the value of interaction is 1 for two agents. Otherwise, the value of interaction is 0 for two agents. We can derive that $v_i^l(j) = v_j^l(i)$ under a cooperative environment.

$$v_i^l(j) = \begin{cases} 1 & \text{If the interaction is successful} \\ 0 & \text{Otherwise} \end{cases} \tag{1}$$

where, successful interaction means that agent a_i and agent a_j have achieved their goals. On the other hand, unsuccessful interaction means that agent a_i and agent a_j have lost their goals.

Definition 2. Trust Model (TM)

$$TM = \langle Ag, \eta_i(j), OT_i(j), P_i(j), ST_i(j), TT_i(j) \rangle,$$

where, agents $a_1, a_2, ..., a_n$ form a set of Ag, $Ag = \{a_1, a_2, ..., a_n\}$.

The method for computing $\eta_i(j)$ is cited from Teacy [5] shown in the followings. $\eta_i(j) \in [0, 1]$ indicates probability of successful interactions between agent i and agent j.

$$\eta_i(j) = \begin{cases} \frac{1+\sum_{l=1}^{s_i^j} v_i^l(j)}{2+s_i^j}, & i \neq j \\ 1, & i = j \end{cases} \tag{2}$$

We denote OT as $OT_i(j)$, which means that agent a_i trusts a_j on a fixed attribute in a fixed situation. If $\eta_i(j) \geq c$, then $OT_i(j) = 1$. Otherwise, $OT_i(j) = 0$. $c \in (0.5, 1]$ is a fixed constant which means the threshold of trust for all agents in Ag.

$$OT_i(j) = \begin{cases} 1 & \eta_i(j) \geq c \\ 0 & \text{Otherwise} \end{cases} \tag{3}$$

Due to $v_i^l(j) = v_j^l(i)$, $s_i^j = s_j^i$, and c is a constant, so $\eta_i(j) = \eta_j(i)$, and $OT_i(j) = OT_j(i)$.

We denote the preference of agent as $P_i(j)$, which means the preference of agent i on agent j. We consider $P_i(j)$ as the threshold of ST based on $\eta_i(j)$.

We denote ST as $ST_i(j)$, which means that agent a_i trusts a_j in a fixed situation.

$$ST_i(j) = \begin{cases} 1 & \eta_i(j) \geq P_i(j) \\ 0 & \text{Otherwise} \end{cases} \tag{4}$$

We denote TT as $TT_i(j)$, which means that agent a_i has no interactions with agent a_j, but agent a_i received a recommendation from agent a_k which has interactions with agent a_j. We illustrate OT, ST and TT in Fig.1.

$$TT_i(j) = \begin{cases} 1 & \{\eta_k(j) \geq P_i(j)\} \wedge \{ST_i(k) = 1\} \\ 0 & \text{Otherwise} \end{cases} \tag{5}$$

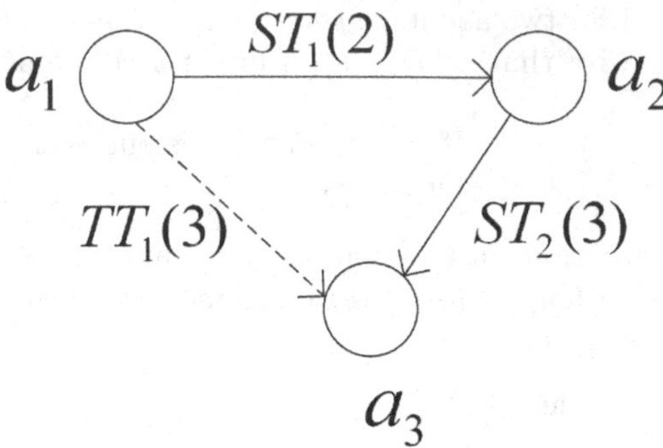

Fig. 1. ST and TT

3 Formation of OT, ST and TT

In this section we will discuss the process of trust formation and some examples are given to understand trust more thoroughly.

3.1 Formation of Trust

Axiom 1. $\forall a_i, a_j \in Ag$, If agent a_i trusts a_j, then agent a_i will delegate his authority to agent a_j. It means that all of interactions did by agent a_j is regarded as agent a_i.

In order to discuss TT, we firstly give some rules of trust shown as follows.

Rule 1. (OT determination rule) $\forall a_i, a_j \in Ag$, If $s_i^j > 0$, and $\eta_i(j) \geq c$, then $OT_i(j) = OT_j(i) = 1$, otherwise $OT_i(j) = OT_j(i) = 0$.

Rule 2. (ST determination rule) $\forall a_i, a_j \in Ag$, If $s_i^j > 0$, and $\eta_i(j) \geq P_i(j)$, then $ST_i(j) = 1$, otherwise $ST_i(j) = 0$.

Rule 3. (TT determination rule) $\forall a_i, a_j, a_k \in Ag$, If agent a_i trusts agent a_j, then interactions between agent a_k and agent a_j are regarded as interactions between agent a_i and agent a_k. The formal description is as follows.

If $s_i^j > 0$, $s_j^k > 0$,$s_i^k = 0$, $ST_i(j) = 1$, and $\eta_j(k) \geq P_i(k)$, then $TT_i(k) = 1$, otherwise $TT_i(k) = 0$.

We consider that agents a_1, a_2, a_3 form trust relationship among them. If $\langle a_1, a_2 \rangle \in ST$, then a_1 regards that the data of a_2 is same to a_1. If agent a_2 recommends a_3 to a_1, then agent $\langle a_2, a_3 \rangle \in ST$, which means that agent a_2 does trust a_3, i.e. $\eta_2(3) \geq P_2(3)$.

However, it does not mean that agent a_1 will trust a_3. Because the data of a_2, integrated with the preference of agent a_2, is transferred to a_1. When a_1 received $ST_2(3)$, it should replace $P_2(3)$ by $P_1(3)$. Agent a_1 derives $TT_1(3)$ based on $P_1(3)$. Therefore, according to Rule 3, the key criterion for the determination of transitivity of ST in this example is whether $\eta_2(3) \geq P_1(3)$.

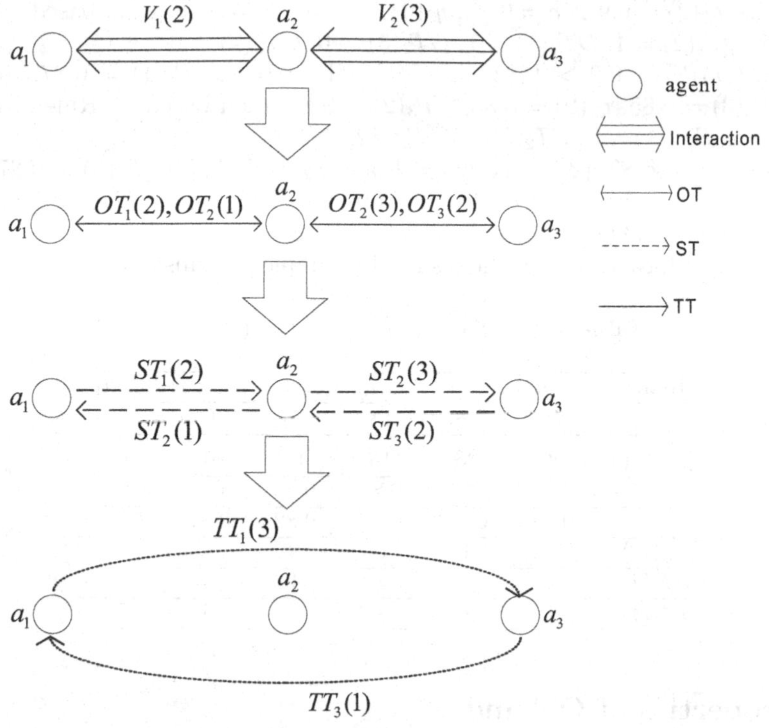

Fig. 2. Formation of OT and ST

Fig.2 describes the formation process of a trust relation. Agents firstly obtain OT based on interactions. Sequentially, agents consider preferences to form ST, and at the end agents should determine whether TT exists among them.

3.2 An Example

Example 1. We discuss trust among buyers and sellers in Amazon for deals of books, and consider that three agents a_1, a_2, a_3 interacted each other. For example, agent a_1 wants to buy a book from agent a_2 in Amazon, because agent a_1 bought lots of books from agent a_2 in Amazon before. But agent a_2 does not have this book, so he recommends another agent a_3 to a_1. Actually, agent a_2 interacted with a_3 before, and agent a_1 did not interact with a_3 in the past. The problem is whether agent a_1 should trust a_3.

We assume that $s_1^2 = s_2^1 = 28$, $\sum_{l=1}^{s_1^2} v_1^l(2) = 26$, $s_2^3 = s_3^2 = 18$, $\sum_{l=1}^{s_2^3} v_2^l(3) = 16$. We also assume that $c = 0.88$, $P_1(2) = 0.88$, $P_2(1) = 0.87$, $P_2(3) = 0.85$, $P_3(2) = 0.82$, $P_1(3) = 0.87$, $P_3(1) = 0.88$.

So according to formula 2, we can get $\eta_1(2) = (26 + 1) \div (28 + 2) = 0.9$, $\eta_2(3) = (16 + 1) \div (18 + 2) = 0.85$, $\eta_1(3) = (0 + 1) \div (0 + 2) = 0.5$.

Because $\eta_1(2) = 0.9 > c = 0.88$, $\eta_2(3) = 0.85 < c = 0.88$, then based on Rule 1, we get $OT_1(2) = 1$, $OT_2(1) = 1$, $OT_2(3) = 0$, $OT_3(2) = 0$.

Because $\eta_1(2) = 0.9 > P_1(2) = 0.88$, $\eta_2(1) = 0.9 > P_2(1) = 0.87$, $\eta_2(3) = 0.85 \geq P_2(3) = 0.85$, $\eta_3(2) = 0.85 \geq P_3(2) = 0.82$, then based on Rule 2, we get $ST_1(2) = 1$, $ST_2(1) = 1$, $ST_2(3) = 1$, $ST_3(2) = 1$.

Finally, because $ST_1(2) = ST_2(3) = 1$, and $\eta_2(3) = 0.85 < P_1(3) = 0.87$, then based on Rule 3, we can get $TT_1(3) = 0$. Moreover, because $ST_3(2) = ST_2(1) = 1$, and $\eta_2(1) = 0.9 > P_3(1) = 0.88$, then based on Rule 3, we can get $TT_3(1) = 1$. The process of trust formation and the result of Example 1 is illustrated in table 1.

Table 1. OT, ST and TT of agents ($c = 0.88$)

Items	$\langle a_1, a_2\rangle$	$\langle a_2, a_1\rangle$	$\langle a_2, a_3\rangle$	$\langle a_3, a_2\rangle$	$\langle a_1, a_3\rangle$	$\langle a_3, a_1\rangle$
s_i^j	28	28	18	18	0	0
$\sum_{l=1}^{s_i^j} v_i^l(j)$	26	26	16	16	0	0
$\eta_i(j)$	0.9	0.9	0.85	0.85	0.5	0.5
$P_i(j)$	0.88	0.87	0.85	0.82	0.87	0.88
$OT_i(j)$	1	1	0	0	0	0
$ST_i(j)$	1	1	1	1	0	0
$TT_i(j)$	-	-	-	-	0	1

4 Properties of OT and ST

In this section, we will propose some useful properties of OT and ST which were not mentioned before. These properties are the base of following conclusions.

4.1 Properties of Trust

Property 1. In the set of Ag, Trust has a property of persistence.

Let $\forall a_i, a_j \in Ag$, if $\langle a_i, a_j\rangle \in OT$, then it is easier to strengthen this relation than destroy it in any situation. We say that this relation once formed, it is stable and becomes firmer and firmer in most situations. Because Once a choice (trust relation) has been made, the truster will tend to seek evidence in favour of this choice.

For a result of interaction, if it is successful, then it will strengthen trust. We also know that the probability of success is about $\eta_i(j)$. On the other hand, if it is not successful, then trust also has much more possibility to be 1 than 0. Because only a result of interaction is not enough to change $\eta_i(j)$ more, and $\eta_i(j) \geq P_i(j)$ is much more possible to be true all along.

Property 2. In the set of Ag, Trust has multi-dimensions.

Let $\forall a_i, a_j \in Ag$, if $\langle a_i, a_j\rangle \in OT$. It means that agent a_i trust a_j on a particular dimension. It does not mean that agent a_i trust all of things of agent a_j. For example, I may trust my brother to drive me to the airport, I most certainly would not trust him to fly the plane!

In this paper we discuss trust on a fixed dimension. Multi-dimensions of trust will be investigated in our another paper.

4.2 Properties of OT

Property 3. In the set of Ag, OT is a reflexive, symmetric, and transitive relation.

Axiom 2. Let $\forall a_i \in Ag$, then $\langle a_i, a_i \rangle \in OT$. In other words, any agent in Ag trusts herself inherently.

Because the interactional statics is symmetric for all agents and the method of computation is same also, $\forall a_i, a_j \in Ag$, if $\langle a_i, a_j \rangle \in OT$, then $\langle a_j, a_i \rangle \in OT$.

Proof: $\forall a_i, a_j \in Ag$, if $\langle a_i, a_j \rangle \in OT$, it means that $\eta_i(j) \geq c$. Because $\eta_i(j) = \eta_j(i)$, so $\eta_j(i) = \eta_i(j) \geq c$. According to definition 2, then $\langle a_j, a_i \rangle \in OT$. **End.**

Suppose any information in Ag is excise and complete, and any agent in Ag is willing to provide her information to others. Thus, $\forall a_i, a_j, a_k \in Ag$, if $\langle a_i, a_j \rangle \in OT$, and $\langle a_j, a_k \rangle \in OT$, then $\langle a_i, a_k \rangle \in OT$.

Proof: $\forall a_i, a_j, a_k \in Ag$, if $\langle a_i, a_j \rangle \in OT$, and $\langle a_j, a_k \rangle \in OT$, it means that $\eta_i(j) \geq c$, and $\eta_j(k) \geq c$. According to Axiom 1, $\eta_j(k) = \eta_i(k) \geq c$, then $\langle a_i, a_k \rangle \in OT$. **End.**

Corollary 4. OT is an equivalence relation.

Proof. With respect to OT, it is a reflexive, symmetric, and transitive relation according to Property 3. So it is an equivalence relation therefore. **End.**

4.3 Properties of ST

Property 5. In the set of Ag, ST is a reflexive, asymmetric, and nontransitive relation.

Axiom 3. Let $\forall a_i \in Ag$, then $\langle a_i, a_i \rangle \in ST$. In other words, any agent in Ag trusts herself inherently.

Although the interactional statics is symmetric for all agents and the method of computation is same also, due to the different preferences of agents, so $\forall a_i, a_j \in Ag$, if $\langle a_i, a_j \rangle \in ST$, then $\langle a_j, a_i \rangle \in ST$ is not always true.

Suppose any information in Ag is excise and complete, and any agent in Ag is willing to provide her information to others. Because of the different preferences of agents. Thus, $\forall a_i, a_j, a_k \in Ag$, if $\langle a_i, a_j \rangle \in ST$, and $\langle a_j, a_k \rangle \in ST$, then $\langle a_i, a_k \rangle \in ST$ is not always true.

According to Rule 3, the key criterion for determination of transitivity of ST is whether $\eta_j(k) \geq P_i(k)$.

5 Background and Related Work

Paul Marsh [4] firstly studied trust by game theory and distributed artificial intelligence. He gave a computational model for trust and put forward some basis properties of trust. Trust is a complex subject relating to belief in honesty, truthfulness, competence, reliability etc. of the trusted person or service.

The mainstream of trust researches is to describe agent trust based on interaction history by probability function, which is OT in nature. In particular, it is well known that Josang and Ismail [7,8] firstly proposed the Beta Reputation System (BRS), which is based on the beta distribution of probability theory. Agents are required to collect interactions data among them, such as success, failure or others. Moreover, agents will give ratings to the performance of other users in the community. Here, ratings consist of a single value that is used to obtain positive and negative feedback values. However, BRS system is specifically designed for online communities and is centralized. Tong [15] took Fuzzy reasoning method to describe agent trust relationship and a long-term coalition system was proposed as a result.

BRS system is totally depended on objective data between agents who interacted with each other in the past. Even so, there are some unfair ratings, either unfairly positive or negative, towards a certain agent. Whitby et al. [9] extended BRS system and show how it can be used to filter unfair ratings. Yu described another method to filter inaccurate reputation [10]. Subsequent observations of trustee behavior are provided to the system as opinion sources. At this point, different methods are adopted to represent trust, ground trust in trustee observations, and implement reputation filtering. Teacy extended BRS and put forward TRAVOS model [5], which can treat with inaccurate information effectively. TRAVOS also provided the confidence of trust computation. If the value of confidence is under minimum predefined value, trust is substituted by reputation. An important advantage is that it can overcome noise and lying sources of reputation effectively. Furthermore, Tong [13] paid more attention to dynamic variety of agent trust for precise prediction and abnormal behavior detection of trust. CMAIT model was proposed based on derivative of trust.

Hang [18] synthesized operators for Propagating Trust in the social networks, such as concatenation, aggregation. A new operator, selection, was suggested to improve the computational system. JensWitkowski and Kastidou [16,17] studied on the honesty of trustees who offered the data of interactions.

Another method is to investigate structure model of trust, which will integrate trust computational model to produce a comprehensive assessment of another agent's performance. Huynh proposed an open multi-agent system, named FIRE model [11], which integrated trust and reputation model. It incorporated interaction trust, role-based trust, witness reputation, and certified reputation to provide a trust metric in most circumstances, where role-based trust is a rule reasoning trust, witness reputation is depended on external observation and certified reputation is computed through third party.

All of the above work are based on probability computational models of trust and never consider subjective component of trust. Particularly, inaccurate information offered by others is partly due to the subjectivity of trust. So an in-depth study is urgently necessary for the subjectivity and objectivity of trust. Tong

[12,14] has investigated ST integrated with OT from agent preferences. Based on probability theory method, a preference is considered to revise the trust.

6 Conclusions

In this paper, we clarified the differences between ST and OT firstly. Particularly, we investigated ST and OT and proposed some useful properties of trust. We discussed the formation of trust and gave an example to understand trust deeply.

For the future work on trust researches, we think that trust is indeed a kind of belief, so trust evaluation naturally can be treated as belief revision. In our another papers we will investigate trust as belief revision.

Acknowledgment. The authors would like to thank School of Computer Science of Yantai University in China. In the laboratory of Intelligence Information Processing, the authors have received much enthusiasm and help from the staff without any personal motivations.

This work is partly supported by the Major State Basic Research Development Program of China under Grant (2007CB307100), the National Natural Science Foundation of China (61170224,60973075), the Shandong Natural Science Foundation (ZR2011FL018,2012GGB01017), and the Technology Research Program of Education Commission of Shandong Province (J11LG35,J10LG27).

References

1. Cao, L.: Data Mining and Multi-agent Integration (edited). Springer (2009)
2. Cao, L., Gorodetsky, V., Mitkas, P.A.: Agent Mining: The Synergy of Agents and Data Mining. IEEE Intelligent Systems 24(3), 64–72 (2009)
3. Cao, L., Weiss, G., Yu, P.S.: A Brief Introduction to Agent Mining. Journal of Autonomous Agents and Multi-Agent Systems 25, 419–424 (2012)
4. Marsh, S.P.: Formalising Trust as a Computational Concept. (Dissertation), University of Stirling (1994)
5. Luke Teacy, W.T., Patel, J., Jennings, N.R., Luck, M.: TRAVOS: Trust and reputation in the context of inaccurate information sources. Journal of Autonomous Agent Multi-Agent System 12, 183–198 (2006)
6. Khosravifar, B.: Maintenance-based Trust for Multi-Agent Systems. In: AAMAS 2009 (2009)
7. Ismail, R., Josang, A.: The beta reputation system. In: Proc. 15th Bled Conf. on Electronic Commerce (2002)
8. Josang, A., Ismail, R., Boyd, C.: A survey of trust and reputation systems for online service provision. Decision Support Systems (2005)
9. Whitby, A., Josang, A., Indulska, J.: Filtering out unfair ratings in bayesian reputation systems. In: Proc. 7th Int. Workshop on Trust in Agent Societies (2004)
10. Yu, B., Singh, M.P.: Detecting deception in reputation management. In: AAMAS 2003, pp. 73–80 (2003)
11. Huynh, T.D., Jennings, N.R., Shadbolt, N.R.: An integrated trust and reputation model for open multi-agent systems. Journal of Autonomous Agent Multi-Agent System 13, 119–154 (2006)

12. Tong, X., Huang, H., Zhang, W.: Agent long-term coalition credit. Expert Systems with Applications 36(5), 9457–9465 (2009)
13. Tong, X., Huang, H., Zhang, W.: Prediction and Abnormal Behavior Detection of Agent Dynamic Interaction Trust. Journal of Computer Research and Development 46(8), 1364–1370 (2009) (in Chinese)
14. Tong, X., Zhang, W.: Group trust and group reputation. In: ICNC 2009 (2009)
15. Tong, X., Zhang, W.: Long-Term MAS Coalition Based on Fuzzy Relation. Journal of Computer Research and Development 43(8), 1445–1449 (2006) (in Chinese)
16. Kastidou, G., Larson, K., Cohen, R.: Exchanging Reputation Information Between Communities: A Payment-Function Approach. In: IJCAI 2009 (2009)
17. Witkowski, J.: Eliciting Honest Reputation Feedback in a Markov Setting. In: IJCAI 2009 (2009)
18. Hang, C.-W., Wang, Y., Singh, M.P.: Operators for Propagating Trust and their Evaluation in Social Networks. In: AAMAS 2009 (2009)

KNN-Based Clustering for Improving Social Recommender Systems*

Rong Pan[1], Peter Dolog[1], and Guandong Xu[2]

[1] Department of Computer Science, Aalborg University, Denmark
{Rpan,Dolog}@cs.aau.dk
[2] Centre for Applied Informatics, Victoria University, Australia
Guandong.Xu@vu.edu.au

Abstract. Clustering is useful in tag based recommenders to reduce sparsity of data and by doing so to improve also accuracy of recommendation. Strategy for the selection of tags for clusters has an impact on the accuracy. In this paper we propose a KNN based approach for ranking tag neighbors for tag selection. We study the approach in comparison to several baselines by using two datasets in different domains. We show, that in both cases the approach outperforms the compared approaches.

Keywords: Tag Neighbors, Clustering, Personalization, Recommender Systems, Social Tagging.

1 Introduction

A large number of social tagging sites, such as Delicious, Last.fm, Flickr, CiteULike and Digg have undergone tremendous growth in the past few years.

Tagging, as a labeling of items with specific lexical information, plays a crucial role in such social collaborative tagging systems. The common usage of tags in these systems is to add the tagging attribute as an additional feature to re-model users or resources over the tag vector space, and in turn, making tag-based collaborative filtering recommendation or personalized recommendation. The user-contributed tags are not only an effective way to facilitate personal organization but also provide a possibility for users to search for needed information.

With the help of tagging data, user annotation preference and document topical tendency are substantially coded into the profiles of users or documents. However, the redundancy, ambiguity and syntactic nature of tags are often incurred in all kinds of social tagging systems and remain still the main problems. These problems also impact on the accuracy of recommendation.

In our previous research [20], we have proposed a clustering algorithm which uses tag neighbors. This has improved the precision of the Recommendations; however, the result of selection of tag neighbors for the individual tag in the real-world dataset is not satisfied. So we aim at improving this selection by ranking tag neighbors.

* This research has been partially supported by the EU FP7 ICT project M-Eco: Medical Ecosystem Personalized Event-based Surveillance (No.247829).

L. Cao et al.: ADMI 2012, LNAI 7607, pp. 115–125, 2013.

Due to above problems our motivation in this paper is to expand the previous work [19,23,20,24] with KNN algorithm in order to enhance recommendation precision. The approach is based on KNN neighbor directed graph of tags which is used to drive the expansion of tag expression with top-K frequently tag neighbors. By doing so we are able to facilitate the organization of information documents in search and navigation.

The main contributions made in this paper are:

- We propose an approach of KNN algorithm with tag clustering algorithm to filter the "noisy tags".
- We show that our approach outperform the baselines in experiments on two real world datasets.

The rest of the paper is organized as follows: Section 2 presents the related work in the field of tag neighbors and tag clustering. In section 3, we discuss the preliminaries of the tagging data model. Section 4 describe the previous work for the detailed process of extending the tag neighbors. The experiment design, evaluation measures and the comparison of the results are in Section 5. We conclude the paper in section 6.

2 Related Work

Tags are used to index, annotate and retrieve resource as an additional metadata of resource. Poor retrieval performance remains a major problem of most social tagging systems resulting from the severe difficulty of ambiguity, redundancy and less semantic nature of tags. Clustering method is a useful tool to address the aforementioned difficulties. We review the related literatures from the perspectives of tag expansion and tag clustering.

The authors in [11] formalize the notion of resource recommendation in social annotation systems. A linear-weighted hybrid framework for making recommendations is proposed and shown to be effective. Some of the hybrid recommender systems [4] have been shown to be an effective method of drawing out the best performance among several independent component algorithms. Our work can be also considered as a hybrid recommender system where we utilize tags to understand both, users and documents.

K. R. Bayyapu and P. Dolog in [1] try to solve the problems of sparse data and low quality of tags from related domains. They suggest using tag neighbors for tag expression expansion. The tag neighbors are based on the location within documents while in this paper the tag neighbors are understood as similarity neighbors in vector space model along users and documents.

Recommenders can assist users by suggesting resources, tags or even other users. Authors in [16] have demonstrated that an integrated approach which exploits all three dimensions of the data (users, resources, tags) perform superior results in tag recommendation. They extend the resource recommendation approach and propose an approach for designing weighted linear hybrid resource recommenders. In this paper, we differ in calculating weights and in a way which we apply clustering.

The authors in [24] focus on the sensitivity of initialization instead and make use of the approximate backbone of tag clustering results to find out better tag clusters. By proposing an APProximate backbone-based Clustering algorithm for Tags (APPECT), they fix the approximate backbone as the initial tag clustering result and then assign the rest of the tags into the corresponding clusters based on the similarity.

Existing data mining tools use workflow to capture user requirements. In agent-based distributed data mining (ADDM), agents are an integral part of the system and can seamlessly incorporate with workflows. The authors [18] describe a mechanism to use workflow in descriptive and executable styles to incorporate between workflow generators and executors. They show that agent-based workflows can improve ADDM interoperability and flexibility, and also demonstrates the concepts and implementation with a supporting the argument, a multi-agent architecture and an agent-based workflow model are demonstrated. And the authors [17] investigate an interaction mechanism between agents and data mining, and focus on agent-enhanced mining. The workflow enactment can be improved with a suitable underlying execution layer, which is a Multi-Agent System (MAS). They propose a strategy to obtain an optimal MAS configuration from a given workflow when resource access restrictions and communication cost constraints are concerned, which is essentially a constraint optimization problem. The authors in [6,5] specify five types of ubiquitous intelligence: data intelligence, human intelligence, domain intelligence, network and web intelligence, organizational intelligence, and social intelligence. They define and illustrate them, and discuss techniques for involving them into agents, data mining, and agent mining for complex problem-solving. Further investigation on involving and synthesizing ubiquitous intelligence into agents, data mining, and agent mining will lead to a disciplinary upgrade from methodological, technical and practical perspectives.

In [10,12,13,7] topic relevant partitions are created by clustering resources rather than tags. By clustering resources, the improvement of recommendations is made by distinguishing between alternative meanings of a query. In [3], an interesting approach is proposed to model the documents in social tagging systems by document graphs. The relevance of tag propagated along edges of the documents graph is determined via a scoring scheme, with which the tag prediction was carried out. In [2], an approach that monitors users activity in a tagging system and dynamically quantifies associations among tags is presented. The associations are then used to create tag clusters. In our work we focus on tag clustering instead.

Zhou et al. propose a novel method to compute the similarity between tag sets and use it as the distance measure to cluster web documents into groups. Major steps in such method include computing a tag similarity matrix with set-based vector space model, smoothing the similarity matrix to obtain a set of linearly independent vectors and compute the tag set similarity based on these vectors. [15]. In this paper we propose a different enhanced approach which utilizes tag neighbors, the KNN graph and clustering to compute recommendations.

The purpose of tag clustering is the ability of aggregating tags into topic domains. [22] demonstrate how tag clusters serving as coherent topics can aid in the social recommendation of search and navigation. They present a personalization algorithm for recommendation in folksonomies which relies on hierarchical tag clusters. Their basic recommendation framework is independent of the clustering method. They use a context-dependent variant of hierarchical agglomerative clustering which takes into account the user's current navigation context in cluster selection. We employ clustering in different stage and utilize KNN graph to calculate neighbors. A framework named Semantic Tag Clustering Search, which is able to cope with the syntactic and semantic tag variations is proposed in [8]. In our work we do not consider the semantics of tags as it is an additional computation step. We show, that even without the consideration of semantics the performance of a recommender is reasonable.

3 Preliminaries

We will review the preliminaries based on our previous work in this section [19,23,20,24].

3.1 Folksonomy

The folksonomy is a three-dimensional data model of social tagging behaviors of users on various documents. It reveals the mutual relationships between these three-fold entities, i.e. user, document and tag. A folksonomy F according to [14] is a tuple $F = (U,\ T,\ D,\ A)$, where U is a set of users, T is a set of tags, D is a set of web documents, and $A \subseteq U \times T \times D$ is a set of annotations. The activity in folksonomy is $t_{ijk} \subseteq \{(u_i, d_j, t_k) : u_i \in U, d_j \in D, t_k \in T\}$, where $U = \{U_1, U_2, \cdots, U_M\}$ is the set of users, $D = \{D_1, D_2, \cdots, D_N\}$ is the set of documents, and $T = \{T_1, T_2, \cdots, T_K\}$ is the set of tags. $t_{ijk} = 1$ if there is an annotation (u_i, d_j, t_k); otherwise $t_{ijk} = 0$.

3.2 User Profile and Document Profile

The constructed folksonomy data model is actually a three-dimensional array (or called tripartite hyper-graph). In real applications, we often decompose the tagging data into two two-dimensional matrices, i.e., user profile and document profile. User profile can be used to store the descriptive tags of the user's characteristics and preferences. The document profile is represented by the tags generated by the group of users tagging the documents. In the context of social tagging systems, the user profiles and document profiles thus are expected to be represented by the representative tags. Therefore the process of user and document profiling is to capture the significant tags from a large volume of tagging data in a social collaborative environment. We will utilize them based on our previous work [19].

3.3 Similarity Measure for Tags

The similarity is a quantity that reflects the strength of relationship between two objects. In the previous part, each user profile and document profile can be represented by the pair of tags and frequencies. In this manner, we perform the transformation of above two matrices and utilize the cosine function to measure the similarity between two tags. Its value ranges from 0 to 1, the higher value of the similarity, the more similar the objects are [19].

4 Tag Clustering with Tag Neighbors

The basic idea of the proposed approach is to utilize the tag neighbors to extend the users' or documents' profiles which are represented by the tags. In the tag similarity matrix, each tag has different similarity weights with other tags, we assume the higher the weight, the more similar the tags are to the target tag. To realize the task of expanding the tag, the major difficulty is how to define the tag neighbors and how to locate them from the total tags. Here we adapt a statistical definition of tag neighbor - the tags which are co-occurred most frequently or they have the higher similarity weight in the similarity matrices to the target tag, are the neighbors for each other. So the N tags according to the top-N similarity weight can be defined as the tag neighbors of an individual tag. After such steps, each tag will have an additional neighboring tag set which will help to improve the quantity of the tag expression.

Given an arbitrary tag T, its neighboring tags are defined as:

$$Nb(T_i) = \{T_j : T_j \in TopN\{SM(T_i, T_j)\}\}$$

where $TopN\{SM(T_i, T_j)\}$ is the tags which possess the top-N highest similarity values to tag T_i.

However, not all social tagging systems proposed so far maintain high quality of tag data. Even when the tag elements in tag expression can be expanded, such tag expression could contain a lot of inappropriate tags. We particularly call them as "noisy tags". From the previous paper, we can get the basic idea of filtering: the neighboring tags from the same tag cluster contribute collaboratively to specific topic, being kept as the appropriate tag neighbors for tag expression expansion. The next processing step is to filter out the noisy tags according to the discovered tag clusters. Each tag has an expanded set of tag neighbors, which can belong to different clusters. To ensure all neighboring tags are from the same tag cluster, each tag in the expanded neighbors set will be compared with all the tags from the tag cluster where the target tag is assigned. If the expanded neighbor appears in the same cluster, it can be considered as the appropriate neighbor of the tag, making it kept in the expanded tag set; otherwise, it should be filtered out. After such steps above, the left elements could be defined as the tag neighbors for the target tag, and the quality of the tag neighbor will be accordingly improved. Also in such way the density in the integrated tag-user-document matrix could be increased substantially. For the detailed process, please refer to our previous work[20].

5 Tag Neighbors Filtering Based on KNN Algorithm

In this section we use the k-nearest neighbor algorithm Algorithm to pre-process the tagging data for clustering to optimize the tag clustering algorithm.

The k-nearest neighbor algorithm (KNN) is a method for classifying objects based on closest training examples in the feature space. It's the Memory-based without the training model where the function is only approximated locally and all computation is deferred until classification. The k-nearest neighbor algorithm can generate the maximum likelihood estimation of the class posterior probabilities.The basic idea is: an object is classified by a majority vote of its neighbors, with the object being assigned to the class most common amongst its k nearest neighbors (k is a positive integer, typically small). If $k = 1$, then the object is simply assigned to the class of its nearest neighbor. The best choice of k depends upon the data; the larger values of k reduce the effect of noise on the classification, but make boundaries between classes less distinct[9,21].

According to subsection of Similarity Measure for Tags, the annotation between the user and resource can create a similarity matrix S to indicate the affinity of tags. The k-nearest neighbors for each tag can be created according to S, then we can construct a KNN directed graph G based on the relationships among the tags. So we can define the KNN directed graph G and its adjacency matrix A as [23]:

$G =< V, E >$, where V is the node set of tags and E is the directed edge set between each pair of tags, $< p, q >\in E$ denotes that tag q is one of the KNN-neighbors of tag p.

The adjacency matrix is defined as A, where $A(p,\ q) = 1$, if the directed arch $< p, q >$ exists, and $A(p,\ q) = 0$, otherwise.

Especially in the context of social tagging systems, we know each tag can be expressed as a column vector of user profile and document profile, i.e., $T_i = TU_i$ and $T_i = TD_i$. Then we stack up TU_i and TD_i and form a new tag vector over users and documents, $T_i = TU_i \cup TD_i$. At last, we employ KNN algorithm and spectral clustering on the integrated tag vectors to get tag clusters. The pseudo codes of the spectral tag clustering are listed in Algorithm 1.

6 Experimental Evaluations

We evaluate our approach through extensive experiments. The experiments are conducted on the real world dataset: "MovieLens" dataset and the "M-Eco system" datasets. Our experiments focus on results conducted on the implemented tag-based recommender system with tag neighbor expansion. The goal of such experiments is to show how can we reduce sparsity of data and improve the accuracy of recommendation tag based recommenders to reduce sparsity.

6.1 Dataset and Experimental Setup

The "MovieLens" dataset is provided by GroupLens. We utilize part of the "MovieLens" dataset, which contains tags provided by users on movies. All users

Algorithm 1. KNN Spectral Tag Clustering

Input: The tag-user matrix and tag-document matrix, i.e., $TU = \{TU_i, i = 1, \cdots, K\}$, $TD = \{TD_i, i = 1, \cdots, K\}$

Output: A list of top-N documents for the candidate user and a set of C tag clusters $TC = \{TC_c, c = 1, \cdots, C\}$ such that the cut of C-partitioning of the bipartite graph is minimized

1 Pre-process the tag-user-document matrix by stacking up the above two matrix, $T_i = TU_i \cup TD_i$ according to the a Folksonomy model;
2 Obtain the matrixes of user profiles or document profiles by making a two-dimensional projection of the URT matrix;
3 Calculate all of cosine similarities to score the resemblance of resources to those found in the user profile and document profile;
4 Construct a KNN directed graph;
5 Calculate the diagonal matrices D of T;
6 Form a new matrix $RT = D^{-1/2}TD^{-1/2}$;
7 Perform SVD (Singular value decomposition) operation on RT, and obtain k singular vectors L_s to create a new projection matrix;
8 Execute a clustering algorithm and return clusters of tags in the integrated vectors: $TC = \{TC_c, c = 1, \cdots C\}$;
9 Collect the top-N tags according to the highest values in the similarity matrix for each tag;
10 Partition tags into different clusters by spectral clustering algorithm;
11 Check all the tag neighbors generated in the previous steps whether they belong to the same cluster with the original tag, and filter out noisy tag neighbors;
12 Update the tag vectors of user profiles and document profiles with tag neighbors;
13 Calculate the similarity between the candidate user's tag vector and the document tag vector, and rank the documents according to the similarity values in a descending order;
14 Select the top-N documents with the first N highest similarities as the recommendions to the candidate user.

selected had rated at least 50 movies with rank from 4. We also involve a part of the "M-Eco system" dataset, which provided from M-Eco project. It has users for each medical condition mapped with tags generated for documents matching to a particular medical condition. The "MovieLens" data includes 97 users, 1201 documents and 1712 tags. And the "M-Eco system" has 35 different users, 101 documents and 619 tags with 2017 tagging annotations. Following the common protocol in information retrieval domain which chooses 20-30% data for the testing data, we use 75% of data as the training data and the remaining 25% as the testing data to evaluate our approach.

6.2 Precision Evaluation

We calculate the precision with the same process in our previous work [20].

In the experiment, we examine the different algorithms' precision of recommending documents to the individual user. We calculate the precision in the following steps: The recommendation can be created by various approaches by ranking the similarity values. Each one can generate the top-N documents which will be recommended to the user. Then we can compare them with the existing documents in the test data. If there k documents appear out of the N recommended documents in the test data, the number of test documents existing for each user is K_i , the precision for the individual user is defined as t:

$$t = \frac{K_i}{N} \times 100\%$$

We assume that the existed documents are the preference of each user. The high value of t the better recommendation we got. There are 307 existing documents in the "MovieLens" data and 36 existing documents in the "M-Eco" data for testing. We calculated the precision for each algorithms to evaluate the approach. So that we discussed different strategies of our approach against existing approaches. The "Pure Tag Approach" is to calculate the directly similarity between the user profile and the document profile in the tag vector, and the system will recommend the N documents to the user according to the top-N similarity values. "TagNeighbor Approach" is to calculate such similarity based on the naive tag neighbor expansion by collaborative filtering approach. "TagNeighbor with Clustering Approach" is to filter the noisy tags based on the "TagNeighbor Approach". The "Collaborative Filtering Approach" is set as the benchmark of the experiment. Our proposed approach "KNN-based Clustering TagNeighbor" is to utilize the KNN algorithm combining with clustering to filter out the noisy tags to improve the accuracy of tag neighbors to get the better performance of recommendation.

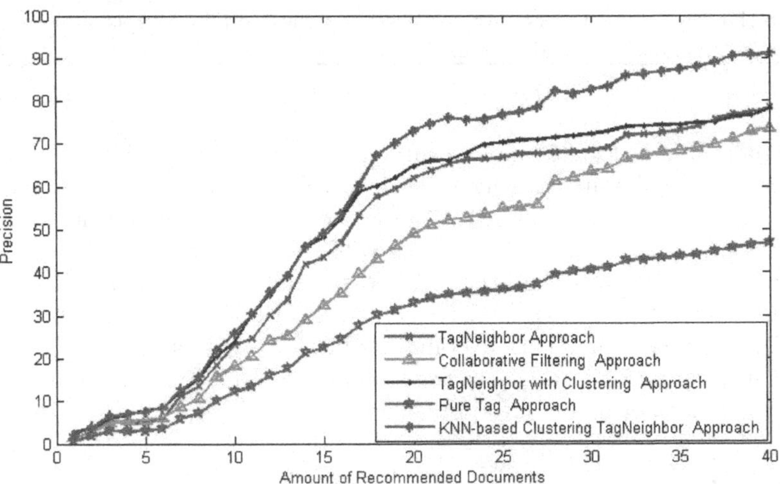

Fig. 1. Precision for conditional recommended documents in "MovieLens"

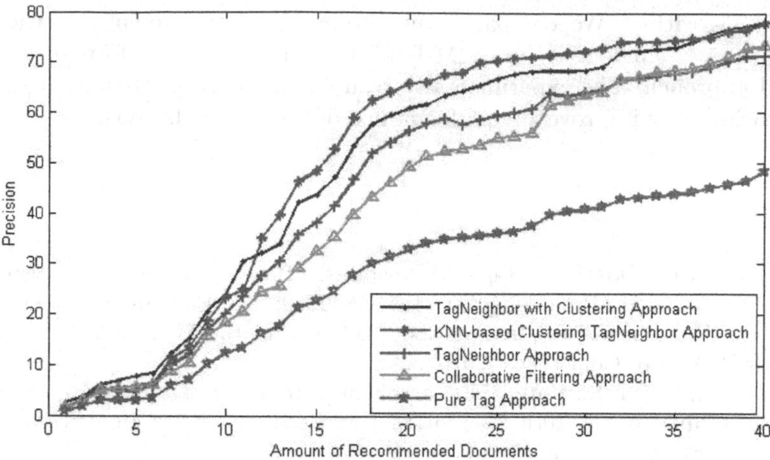

Fig. 2. Precision for conditional recommended documents in "M-Eco"

We average the whole precision for all of the users, and compare the recommendations from top 1 to 40 documents to the users in the experiments. We denote these five approaches as "KNN-based Clustering TagNeighbor", "TagNeighbor with Clustering", "TagNeighbor", "Collaborative Filtering" and the "Pure Tag". For the "M-Eco" dataset, the average precision of "KNN-based Clustering TagNeighbor" is 73.9, "TagNeighbor with Clustering" is 69.1, the "TagNeighbor" is 65.3, the "Collaborative Filtering" is 51.9, and the "Pure Tag" is 29.4. For the "Moivelens system" dateset, the average precision of "KNN-based Clustering TagNeighbor" is 87.1, "TagNeighbor with Clustering" is 82.4, the "TagNeighbor" is 73.3, the "Collaborative Filtering" is 62.1, and the "Pure Tag" is 37.9. The precision comparisons of five approaches for the top 40 recommendations in two real world dataset are shown in Fig. 1 and Fig. 2. In summary, the evidence demonstrates the advantage of our approach in recommendations.

7 Conclusion

Social annotations systems enable users to annotate resources with tags. Under social tagging systems, a typical Web2.0 application, users label digital data sources by using tags which are freely chosen textual descriptions. Tags are used as one kind of specific lexical information that is user-generated metadata with uncontrolled vocabulary, plays a crucial role in such social collaborative tagging systems.

In this work, We aim on the major problem of most social tagging systems resulting from the severe difficulty of ambiguity, redundancy and less semantic nature of tags, we have proposed a method to utilize the KNN algorithm combining with the clustering approach for expanding tag neighbors and improve the accuracy of recommendation. We improved our previous work by involving the KNN algorithm. We compared the preliminary experiments on the real-world dataset: "MovieLens" and "M-Eco" to evaluate the performance of our proposed approach. The experimentally result demonstrates that our approach could considerably improve the performance of recommendations.

References

1. Bayyapu, K.R., Dolog, P.: Tag and Neighbour Based Recommender System for Medical Events. In: Proceedings of MEDEX 2010: The First International Workshop on Web Science and Information Exchange in the Medical Web Colocated with WWW 2010 Conference (2010)
2. Boratto, L., Carta, S., Ratc, V.E.: A robust automated tag clustering technique. In: Proceedings of the 10th International Proceedings on E-Commerce and Web Technologies, pp. 324–335 (2009)
3. Budura, A., Michel, S., Cudré-Mauroux, P., Aberer, K.: Neighborhood-Based Tag Prediction. In: Aroyo, L., Traverso, P., Ciravegna, F., Cimiano, P., Heath, T., Hyvönen, E., Mizoguchi, R., Oren, E., Sabou, M., Simperl, E. (eds.) ESWC 2009. LNCS, vol. 5554, pp. 608–622. Springer, Heidelberg (2009)
4. Burke, R.: Hybrid recommender systems: Survey and experiments. In: User Modeling and User Adapted Interaction, pp. 331–370. Springer, Heidelberg (2002)
5. Cao, L., Gorodetsky, V., Mitkas, P.: Agent mining: The synergy of agents and data mining. IEEE Intelligent Systems 24(3), 64–72 (2009)
6. Cao, L., Luo, D., Zhang, C.: Ubiquitous Intelligence in Agent Mining. In: Cao, L., Gorodetsky, V., Liu, J., Weiss, G., Yu, P.S. (eds.) ADMI 2009. LNCS, vol. 5680, pp. 23–35. Springer, Heidelberg (2009)
7. Chen, H., Dumais, S.: Bringing order to the web: automatically categorizing search results. In: CHI 2000: Proceedings of the SIGCHI Conference on Human Factors in Computing Systems, pp. 145–152. ACM, New York (2000)
8. van Dam, J.-W., Vandic, D., Hogenboom, F., Frasincar, F.: Searching and browsing tag spaces using the semantic tag clustering search framework. In: Proceedings of the 2010 IEEE Fourth International Conference on Semantic Computing, ICSC 2010, pp. 436–439. IEEE Computer Society, Washington, DC (2010)
9. Dasarathy, B.V.: Nearest Neighbor (NN) Norms: NN Pattern Classification Techniques (1991)
10. Di Matteo, N.R., Peroni, S., Tamburini, F., Vitali, F.: A parametric architecture for tags clustering in folksonomic search engines. In: Proceedings of the 2009 Ninth International Conference on Intelligent Systems Design and Applications, ISDA 2009, pp. 279–282. IEEE Computer Society, Washington, DC (2009)
11. Gemmell, J., Schimoler, T., Mobasher, B., Burke, R.: Tag-Based Resource Recommendation in Social Annotation Applications. In: Konstan, J.A., Conejo, R., Marzo, J.L., Oliver, N. (eds.) UMAP 2011. LNCS, vol. 6787, pp. 111–122. Springer, Heidelberg (2011)
12. Guan, Z., Wang, C., Bu, J., Chen, C., Yang, K., Cai, D., He, X.: Document recommendation in social tagging services. In: Proceedings of the 19th International Conference on World Wide Web, WWW 2010, pp. 391–400. ACM, New York (2010)
13. Hayes, C., Avesani, P.: Using tags and clustering to identify topic-relevant blogs. In: International Conference on Weblogs and Social Media (March 2007)

14. Hotho, A., Jäschke, R., Schmitz, C., Stumme, G.: Folkrank: A ranking algorithm for folksonomies. In: LWA, pp. 111–114 (2006)
15. Zhou, J., Nie, X., Qin, L., Zhu, J.: Journal of Computers
16. Gemmell, J., Schimoler, T., Mobasher, B., Burke, R.: Resource Recommendation in Collaborative Tagging Applications. In: Buccafurri, F., Semeraro, G. (eds.) EC-Web 2010. LNBIP, vol. 61, pp. 1–12. Springer, Heidelberg (2010)
17. Moemeng, C., Wang, C., Cao, L.: Obtaining an Optimal MAS Configuration for Agent-Enhanced Mining Using Constraint Optimization. In: Cao, L., Bazzan, A.L.C., Symeonidis, A.L., Gorodetsky, V.I., Weiss, G., Yu, P.S. (eds.) ADMI 2011. LNCS, vol. 7103, pp. 46–57. Springer, Heidelberg (2012)
18. Moemeng, C., Zhu, X., Cao, L.: Integrating Workflow into Agent-Based Distributed Data Mining Systems. In: Cao, L., Bazzan, A.L.C., Gorodetsky, V., Mitkas, P.A., Weiss, G., Yu, P.S. (eds.) ADMI 2010. LNCS, vol. 5980, pp. 4–15. Springer, Heidelberg (2010)
19. Pan, R., Xu, G., Dolog, P.: User and Document Group Approach of Clustering in Tagging Systems. In: Proceeding of the 18th Intl. Workshop on Personalization and Recommendation on the Web and Beyond, LWA 2010 (2010)
20. Pan, R., Xu, G., Dolog, P.: Improving recommendations in tag-based systems with spectral clustering of tag neighbors. In: Proceedings of The 3rd FTRA International Conference on Computer Science and its Applications (CSA 2011): Computer Science and Convergence. LNEE, vol. 114, Part I, pp. 355–364. Springer, Heidelberg (2011)
21. Shakhnarovish, D., Indyk: Nearest-Neighbor Methods in Learning and Vision. The MIT Press (2005)
22. Shepitsen, A., Gemmell, J., Mobasher, B., Burke, R.: Personalized recommendation in social tagging systems using hierarchical clustering. In: Proceedings of the 2008 ACM Conference on Recommender Systems, RecSys 2008, pp. 259–266. ACM, New York (2008)
23. Xu, G., Zong, Y., Pan, R., Dolog, P., Jin, P.: On Kernel Information Propagation for Tag Clustering in Social Annotation Systems. In: König, A., Dengel, A., Hinkelmann, K., Kise, K., Howlett, R.J., Jain, L.C. (eds.) KES 2011, Part II. LNCS, vol. 6882, pp. 505–514. Springer, Heidelberg (2011)
24. Zong, Y., Xu, G., Jin, P., Zhang, Y., Chen, E., Pan, R.: APPECT: An Approximate Backbone-Based Clustering Algorithm for Tags. In: Tang, J., King, I., Chen, L., Wang, J. (eds.) ADMA 2011, Part I. LNCS, vol. 7120, pp. 175–189. Springer, Heidelberg (2011)

A Probabilistic Model Based on Uncertainty for Data Clustering

Yaxin Yu[1], Xinhua Zhu[2], Miao Li[1], Guoren Wang[1], and Dan Luo[2]

[1] College of Information Science and Engineering, Northeastern University, China
{Yuyx,Wanggr}@mail.neu.edu.cn
[2] QCIS, University of Technology, Sydney, Australia
{Xinhua.Zhu,Dan.Luo}@uts.edu.au

Abstract. Recently, all kinds of data in real-life have exploded in an unbelievable way. In order to manage these data, dataspace has been becoming a universal platform, which contains various kinds of data, such as unstructured data, semi-structured data and structured data. But how to cluster these data in dataspace in an efficient and accurate way to help the user manage and explore them is still an intractable problem. In the previous work, the uncertain relationship between term and topic is not considered sufficiently. There are many techniques to handle this problem and probability theory provides an effective way to deal with the uncertainty of clustering. As a result, we proposed a novel probability model based on topic terms, i.e., Probabilistic Term Similarity Model (PTSM) to tackle the uncertainty between term and topic. In this model, not only terms from various data but also structure information of semi-structured and structured data are considered. Each term is assigned a probability indicating how relevant it is to the topic. Then, according to the probability for each term, a probabilistic matrix is established for clustering various data. At last, extensive experiment results show that the clustering method based on this probabilistic model has excellent performance and outperforms some other classical algorithms.

Keywords: uncertainty, probability, topic, data clustering, dataspace.

1 Introduction

Recently, all kinds of data in real-life have exploded in an unbelievable way. In order to manage these data, dataspace has been becoming a universal platform. But how to cluster these data in dataspace in an efficient and accurate way to help the user manage and explore them is still an intractable problem for the researchers.

Although data clustering has been widely researched, most of the previous clustering algorithms express data topic with a certain term set, when analyze the relationships between terms and the topics of data. In other words, terms in the set are invariable. But in fact, the relationships between some special terms and the topic are uncertain. For example, a term t may appear in a datum several times, but it would not certainly be relevant to topic of the datum,

L. Cao et al.: ADMI 2012, LNAI 7607, pp. 126–138, 2013.

maybe is only a common term appearing in many objects. Thus, it is incorrect that we recognize it as an important term to the topic or not. Motivated by this, each term is assigned a probability in this paper, indicating how associated it is with the topic. And then, a topic has several term sets and each of them has a probability. As a result, uncertainty during clustering is solved elegantly.

Furthermore, most existing algorithms only consider the direct similarity relationships between two data, but do not discuss the effect of the transitive relation among the datum similarity to the clustering. Shown as the figure 1, assume a, b, and c are three objects (means data in dataspace), the similarity between a, b and b, c are $s_{a,b}$ and $s_{b,c}$ respectively. It is not reasonable to affirm that there is no relationship between a and c, because they can be communicated indirectly through b. Aiming at this issue, we introduce the concept "hop" to address similarity calculation between any two indirect objects. So the similarity between a and c is obtained by $s_{a,b} \cdot s_{b,c}$ in two-hop case. And this indirect calculation method can be extended to multi-hop condition.

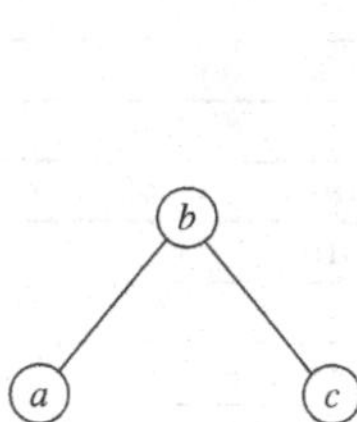

Fig. 1. Indirect similarity relationships

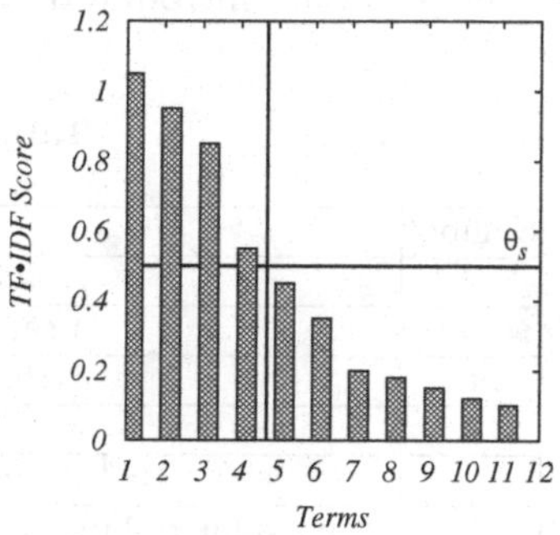

Fig. 2. *TF·IDF* Score Histogram

In this paper, we put forward a novel data clustering model, called Probabilistic Term Similarity Model, for different kinds of data. First, we extract some structured information from semi-structured and structured data, for example, author, keywords, title and table name. And then, give each term a score, according to *TF·IDF* rules. The extracted structured information is assigned a high score. During clustering, these structure information will be treated as a normal term, but has a much higher weight, which means it would obtain a higher probability to indicate the topic of the data. Next, the data are represented in several forms and each of them has a probability to indicate the topic of data. The similarity of any two data is calculated, depending on these different representations. Finally, cluster the data according to a novel similarity matrix consisting of term similarities. Thus, all kinds of data, including un-structured, semi-structured and structured data, will be clustered. The main contributions of this paper are summarized as follows:

1. The model takes the uncertainty between terms and the topic into account and describes it in a probabilistic way.

2. A novel probabilistic similarity matrix is proposed. This matrix not only stores direct relationships between any two data, but also stores indirect relationships between them.
3. The extensive experiments show that the proposed model performs both efficiently and accurately on clustering various kinds of objects in dataspace.

The rest of this paper is organized as follows. Section 2 introduces some preliminaries of data clustering and our approach. The details of our clustering model and the clustering processing are discussed in Section 3. In Section 4, the performance study is described. Section 5 gives some related work, while Section 6 concludes the paper and suggests the future work.

2 Preliminaries

Before we introduce some preliminaries of the clustering processing, the symbols used in this paper are introduced in the Table 1.

Table 1. Notation

Symbol	Definition
d	a datum in dataspace
D	a set used to represent data
t	term appearing in a datum
s	Semi-related term
$r[i]$	i^{th} term in a list of ranked terms
k	rank value of largest plunge of an adjacent ranked term pair
$\sigma_r[i]$	*TF·IDF* score of ith term in a list of ranked terms
θ_u	the threshold between un-related term and semi-related term
θ_s	the threshold between semi-related term and related term
θ_{sim}	the threshold of calculating similarity of two data
θ_c	the threshold of clustering
f_t	the frequency of term t appearing in data d
$p(t)$	topic relevant probability
$s(t)$	*TF·IDF* score of term t
d_i	the ith representation of d
r_i	i^{th} related term
$p(d_i)$	the probability of d_i
M^n	n-connection matrix
M_c	mesh matrix for clustering

As we know, an unstructured datum d could be expressed by a set $d(f_{t_1}, f_{t_2}, ..., f_{t_N})$, where t_i is a term appearing in d, $f_{t_i}(i = 1, 2, ..., N)$ is the frequency of t_i and N is the term size of the whole dictionary. If t_i does not appear in d, f_{t_i} is 0. But for semi-structured data and structured data, expressing data in this way will lose some necessary structure information. Whereas, we still express data in this form here and the structure information would be considered later on. And then, each term is assigned a *TF·IDF* score [1] as the weight during clustering processing.

Therefore, after the *TF·IDF* step, d could be described by $d(s(t_1), s(t_2), ..., s(t_N))$, in which $s(t_i)(i = 1, 2, ..., N)$ means the *TF·IDF* weight to the term t_i for d. The reason why *TF·IDF* is used as the weight of each term is because it has the following characteristics [2], which is appropriate in clustering:

1. Less scores to terms that appear in many data.
2. More scores to terms that appear many times in one datum.
3. Less scores to data that contain many terms.
4. It takes all the data into account, but not only one single datum.

In clustering, we usually want to cluster data having the same topic into a cluster, so topics can be treated as a clustering standard. From the above properties of *TF·IDF*, we can conclude that terms with higher scores are more relative to the topic of data, and terms with lower scores are negligible in clustering, because they are irrelative to the topic of data. So it is reasonable that we get the idea of classifying the terms in a datum into several types, according to the *TF·IDF* score. For example, terms with high scores are classified into a group and low score terms are put into another group. When clustering data, we use terms in different groups to express the topic of data. Next, we will introduce several definitions to describe these different kinds of terms, which play different effects in clustering.

Definition 1. *Unrelated Term is a term whose TF·IDF score is less than the threshold* θ_u.

Definition 2. *Semi-related Term is a term whose TF·IDF score is less than the threshold* θ_s, *but is larger than* θ_u.

Definition 3. *Related Term is a term whose TF·IDF score is larger than* θ_s.

Definition 4. *Topic Term is a term that must be relative to the topic of data.*

As mentioned above, terms with low *TF·IDF* score may be irrelevant to the topic of data, so it would not be a Topic Term. On the contrary, terms with high score may be a Topic Term. So we assume that Unrelated Terms have no relationship with the topic of data, which means they are not Topic Terms, but all Related Terms are Topic Terms, because they get high *TF·IDF* scores. However, whether Semi-related Terms are Topic Terms or not is uncertain. We can give each Semi-related Terms a probability, indicating that the term may be a Topic Term in some probabilistic degree. The probability is called "Topic Relevant Probability" and its definition is given in definition 5.

Definition 5. *Topic Relevant Probability is the probability that a term could be a Topic Term.*

According to definition 5, we can conclude that the topic relevant probability of Unrelated Term is 0, that of Related Term is 1, and Semi-related Term's is between 0 and 1.

3 A Probabilistic Term Similarity Model

In this section, we will introduce our proposed model in details. First of all, we extract structure information such as author, title and attribute name from semi-structured and structured data, which facilitates dealing with various kinds of data. The rest content of the semi-structured and structured data and all the unstructured data are depicted with a term set $d(f_{t_1}, f_{t_2}, ..., f_{t_N})$. Then, the structure information is treated as normal term and put into the set d. Next, according to the possible terms in set d, the data will be expressed with several representations and the similarities between any two data are calculated, depending on these different data expressing. Finally, a novel probabilistic term similarity matrix is established to estimate the comprehensive similarity of data.

3.1 Term Classification

As mentioned above, terms in data are classified into several groups and each term in these groups has a different Topic Relevant Probability. But how to determine the Topic Relevant Probability of these terms is unsettled. So what we need to do first is dealing with terms in different groups, i.e. giving an appropriate probability to terms that can indicate the data topic (Related Term and Semi-related Term) and ignoring the unnecessary terms (Unrelated Term).

First, we rank the terms appearing the data d, according to their values of *TF·IDF* scores in $d(s(t_1), s(t_2), ..., s(t_N))$ and the results are shown in a new kinds of histogram called *TF·IDF* Score Histogram (TISH), which displays the trend of *TF·IDF* values and is shown as Figure 2. The TISH of terms appearing in a datum d represents the distribution of *TF·IDF* score of d, in which, the x-axis represents the rank of those terms appearing in the data and the y-axis represents their corresponding *TF·IDF* scores. Let $\sigma_r[i]$ denote the *TF·IDF* score of term t_i of data d in TISH. Shown in Figure 2, ranked terms $r[1]$, $r[2]$, $r[3]$ and $r[4]$ are grouped into Related Terms. And then, $\sigma_r[4]$ indicates the upper bound of θ_s, where θ_s is the threshold of Related Term. Depending on Definition 3, the Related Terms must have relationships with the topic of data, thus all of the Topic Relevant Probabilities of them are 1.

After extracting Related Terms, we can ignore these terms from TISH. As shown in Figure 2, terms $r[1]$, $r[2]$, $r[3]$ and $r[4]$ can be ignored. The main purpose of ignoring the related terms from TISH is facilitating extracting the semi-related term. In order to achieve this, a heuristic method on TISH is applied for determining the threshold θ_u of Semi-related Term. Let $\theta_u = \sigma_r[v]$, only if $\sigma_r[v] - \sigma_r[v+1] = max_{1 \leq i \leq l-1}(\sigma_r[i] - \sigma_r[i+1])$, where l is the number of terms in ranked term list. Here, θ_u is the *TF·IDF* score of term $r[v]$ whose difference to that of $r[v+1]$ is largest in the ranked term list. That means θ_u is the largest plunge occurs in TISH and indicates the range of Semi-related Term. Referring to Figure 2, the largest plunge of TISH occurs from $r[6]$ to $r[7]$, i.e., $\theta_u = \sigma_r[6]$, and the ranked terms, $r[5]$ and $r[6]$, are grouped into Semi-related Terms. Intuitively, such extraction rule guarantees the most significant terms in the rest terms are extracted as Semi-related Terms for clustering. For any Semi-related Term s, its topic relevant probability $p(s)$ must be between 0 and 1. In this paper, it is set to be $^{s(s)}/_{\theta_s}$.

The purpose of the heuristic method is to find the largest plunge of TISH, which is the largest difference between the two adjacent ranked terms. Through the analysis of a large amount of data, we found that the trend of the plunge between two adjacent ranked terms is degressive from the beginning to the end of TISH. Thus, we could guarantee that the largest plunge of TISH occurs at the forepart of TISH, and the number of the Semi-related Term is not too large. Not using this rule to extract Related Term is because sufficient terms are needed to express the topic of the data.

The rest of the terms are recognized as Unrelated Terms, which are irrelevant with the topic of the data, and could be ignored. In the clustering processing, we do not take them into account. In fact, the main purpose of this step is to reduce the dimensionality of the data representation. And it also can make it easy to represent the topic of data, because after the Unrelated Term is neglected, we can put our attention to those more important terms to judge whether they could indicate the topic of data. Now, any datum could be described by a new set of pairs $D < w_1, w_2, ..., w_n >$. Each element, w_i, in D, is a $< t, p(t) >$ pair, t means the term in the data, which is a Semi-related Term or a Related Term and $p(t)$ indicates the Topic Relevant Probability of the term to the data.

3.2 Probabilistic Representation of Data

If two pieces of data have the same topic, they can be put into the same cluster. But topic is a semantic conception, which is not very easy to be described. Thus we can not express the topic directly and neither do not know which data could be put into a cluster. However, Related term and Semi-related term may be helpful to indicate the topic of the data in an indirect way, because they all have a probability to indicate the datum topic. Next, we will introduce how the topic could be expressed by Related terms and Semi-related terms.

At the beginning of the clustering processing, structure information is extracted from semi-structured and structured data, which is helpful to express the datum topic. Now we can use them for clustering. As mentioned above, they are relevant to the datum topic, so they can be treated as Related Terms. Thus the topic of a datum could be expressed by a new set of terms $d(r_1, r_2, ..., r_n)$, in which r_i means i^{th} related term.

For any Semi-related Term s, it may be a topic term with its Topic Relevant Probability $p(s)$, or not with the probability $1 - p(s)$. Thus we can put the term into term set d with probability $p(s)$, because this term could express the topic of the datum with $p(s)$ and do not put it into d with probability $1 - p(s)$. Then, the datum would be expressed with two ways, $d_1(r_1, r_2, ..., r_n, s)$ with probability $p(s)$ and $d_2(r_1, r_2, ..., r_n)$ with probability $1 - p(s)$. If one datum d has m semi-related terms, it would be expressed in 2^m forms and each representation indicates the datum topic with a probability. Then the probability of each representation to indicate the data topic is $P(d_e) = \prod_{i=1}^{m} P_i$, in which $e = 1, 2, ...2^m$

and d_e means the eth expressing forms. If s_i appears in d_e, $P_i = p(s_i)$, otherwise $P_i = 1 - p(s_i)$.

The next example would illustrate our ideas. Assume there are two pieces of data, d and d'. Data d has 1 Semi-related Term, s, and d' has 2 Semi-related Terms, s'_1 and s'_2. Now d could be expressed in two forms, $d_1(r_1, r_2, ..., r_n, s)$ and $d_2(r_1, r_2, ..., r_n)$. Because s has a probability $p(s)$ to indicate the topic of d, the probabilities of d_1 and d_2 are $p(s)$ and $1 - p(s)$. Similarly, d' could be expressed in four forms $d'_1(r_1, r_2, ..., r_{n'})$, $d'_2(r_1, r_2, ..., r_{n'}, s'_1)$, $d'_3(r_1, r_2, ..., r_{n'}, s'_2)$ and $d'_4(r_1, r_2, ..., r_{n'}, s'_1, s'_2)$, with probabilities $(1 - p(s'_1))(1 - p(s'_2))$, $p(s'_1)(1 - p(s'_2))$, $(1 - p(s'_1))p(s'_2)$ and $p(s'_1)p(s'_2)$.

3.3 Probabilistic Matrix of Term Similarity

After data are transformed to these probabilistic representations, the similarities of all the data in the dataspace are calculated. Fist, the similarity between any two representations of different data is calculated and it also has a probability, which is the product of the probabilities of the two representations. The similarity between any two representations of two different data is defined as follows.

Let d_i be i^{th} representation of d and d'_j be j^{th} representation of d', the similarity of the two representations is:

$$sim(d_i, d'_j) = \frac{|d_i \cap d'_j|}{|d_i \cup d'_j|}$$

Let d have m semi-related terms and d' have m' semi-related terms, and then we have $2^{m+m'}$ similarities between the two data and the probability of each of them equals to the product of the probabilities of the two representations. We can prove that the sum of these similarities' probabilities equals to 1 (it is a very simple mathematic problem, so we do not prove it here).

All similarities of the different representations of d and d' discussed in the examples of Section 3.2 and their probabilities are shown as follows. In that example, there are $2^1 \cdot 2^2 = 8$ similarities between d and d'.

$$\begin{aligned}
&sim(d_1, d'_1) --- p(s)(1 - p(s'_1))(1 - p(s'_2)) \\
&sim(d_2, d'_1) --- (1 - p(s))(1 - p(s'_1))(1 - p(s'_2)) \\
&sim(d_1, d'_2) --- p(s)p(s'_1)(1 - p(s'_2)) \\
&sim(d_2, d'_2) --- (1 - p(s))p(s'_1)(1 - p(s'_2)) \\
&sim(d_1, d'_3) --- p(s)(1 - p(s'_1))p(s'_2) \\
&sim(d_2, d'_3) --- (1 - p(s))((1 - p(s'_1))p(s'_2) \\
&sim(d_1, d'_4) --- p(s)p(s'_1)p(s'_2) \\
&sim(d_2, d'_4) --- (1 - p(s))p(s'_1)p(s'_2)
\end{aligned}$$

After these similarities are calculated, we sum all the probabilities of the similarities whose values are larger than the threshold θ_{sim} together, where θ_{sim} is the threshold of calculating similarity of two data. Let the summation be

$P(d, d')$, which means that d and d' have the same topic with a probability $P(d, d')$. In the above example, if $sim(d_2, d_1')$ and $sim(d_1, d_4')$ are larger than θ_{sim}, $P(d, d') = (1 - p(s))(1 - p(s_1'))(1 - p(s_2')) + p(s)p(s_1')p(s_2')$. Now all the similarity probabilities of data are calculated and could be expressed in a $n \times n$ matrix M, whose elements are $P(d, d')$, d and d' are two data in dataspace.

But this matrix only describes the direct relationship between two data. As mentioned in Section 1, only considering direct relationship of two objects is not sufficient for calculating the similarity of the two objects. The indirect relationships should also be taken into account. The relationships between any two pieces of data can be expressed by a matrix or a graph. In order to simplify the expression of the problem, both the matrix and the graph would be mentioned in the following.

Shown as the Figure 3, assume a, b, and c are three persons, the probability that a can directly communicate with c is $p(a, c)$, and the probabilities between a , b and b, c are $p(a, b)$ and $p(b, c)$ respectively. It is not complete that only consider the probability of direct relationship between a and c, because a could also communicate with c through b. In this case, the probability is $p(a, b)p(b, c)$. Thus, under both direct and indirect paths, the probability that a could communicate with c is $1 - (1 - p(a, b)p(b, c))(1 - p(a, c))$, which means 1 minus the failure probabilities of both two ways.

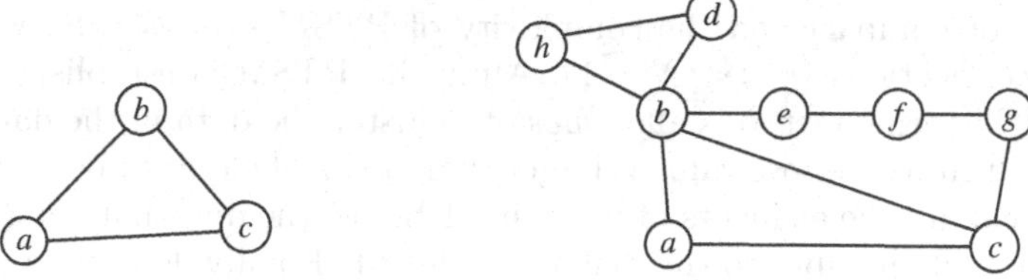

Fig. 3. Relationships among objects **Fig. 4.** An example of n-connection probability

There are many connected paths including one-*hop*, two-*hop*,...,n-*hop* between two objects. Aiming at some hop, for example, k-*hop*, the probability of relationships is $1 - \sum_{i=0}^{m} (1 - p_i)$, in which p_i is the probability of the ith condition of k-*hop*. The next definition is used to calculate p_i.

Definition 6. *Given a graph $G = (V, E)$, V and E are two sets that represent nodes and edges of G respectively. Each edge has a probability that indicates two nodes have some relationship or not. Let p and q be two nodes in G. If there exists n-hop path from p to q, containing no acylic path, in other words, there are no repeated node in the path, the product of all the probabilities of the edges in this path is called n-connection Probability.*

As shown in Figure 4 , let $n = 5$, there would be two paths from a to c with five hops: one is $a \to b \to e \to f \to g \to c$; the other is $a \to b \to h \to d \to b \to c$. But the second path contains an acylic path where node b is repeated, so it is unnecessary, since it equals to the path $a \to b \to c$.

Theorem 1. *Given a graph G, it has N different nodes, there are at most (N-1)-connection Probability between any two nodes in G.*

Proof. Assume that there exists a N-connection probability between two nodes p and q. Then there must be $N+1$ different nodes in the path, which is contrary to the given condition that the graph only has N different nodes.

Definition 7. *Given a graph G, n-connection Failure Matrix, denoted by M^n, is a matrix whose element $M^n_{i,j} = \sum_{h=1}^{A^n_{N-2}} (1 - p^h)$, p^h is the hth n-connection probability between any two nodes in G.*

The n-connection failure matrix stores the probability that all the *n-hop* conditions fails.

Definition 8. *Given a graph G, there are N nodes in the graph. Complete Relationship Matrix, denoted as M_c, is a matrix in which each element $g_{i,j}$ equals to $1 - \prod_{n=1}^{N-1} M^n_{i,j}$ and $M^n_{i,j}$ is the value of the elements of the ith row and j^{th} collum in k-connection failure matrix.*

Complete relationship matrix shows the probabilities of the relationships between any two nodes in G, because it takes all the conditions into account, from 1 hop to $N-1$ hops. In fact, considering all the conditions is meaningless. While the count of hops increases, the connection between the two vertexes of the path becomes not compact and the complexity of PTSM increases. So we only considers k-connections ($k << N-1$). After the PTSM is established, we use a common clustering method called *mesh* to cluster the data in the dataspace. Set the element in M_c, whose value is larger than a threshold θ_c to be 1, otherwise to be 0. We call all the elements of the value 1 below the diagonal, *knotnode*. Then assign the node number to the diagonal element. Finally, longitude (transverse) and woof (vertical line) is drawn from the knot node to the diagonal elements. All the diagonal elements which are on the longitude or woof with the same knot node are clustered into one cluster.

4 Experimental Results

To show the effectiveness and efficiency of our proposed model, experiments on a well-known data set, *NSF Research Awards Abstracts 1990-2003*, are performed. The data of *NSF Research Awards Abstracts 1990-2003* are all unstructured data, i.e. txt files, but each file contains some other information, just like abstract, author, title, time, NSF Org, which could be treated as structured data. We randomly extract 500, 1000, 1500, 2000 papers out of the whole data sets as our experimental data set. Each of them totally includes 30797 terms in the dictionary. And all the experiments are done with Intel Pentium Dual E2200 processor (2.2GHz CPU with 1GB RAM).

There are many different quality measures [3] and the performance of different clustering algorithms can vary substantially depending on which measure is used. So in our experiments, we use F-Measure, entropy, and NMI as the standards to measure the effectiveness performance of the algorithm of PTSM,comparing with CP [4] and K-means. Execution time is used to measure the efficiency of all these algorithms and the effects of parameters of PTSM are also discussed.

4.1 Model Effectiveness

The first quality measure we use to estimate the performance of the algorithm is the F-measure [5], a measure that combines both the precision and recall from information retrieval [6, 7]. Figure 5 shows the performance of our proposed algorithm on F-measure with data sets of different sizes. We use entropy as another measure of quality of the clustering algorithm, whose results is shown in Figure 6. Less entropy means the information of the system is more regular and the information it contains is plentiful, so the value of the entropy of the best clustering algorithm would be 0. Normalized mutual information (NMI) is another widely used performance evaluation measure for determining the quality of clusters [8]. The performances of our proposed algorithm, CP and K-means on NMI are shown in Figure 7 with data sets of different sizes.

From Figure 5, 6 and 7, it is obvious that our algorithm is better than CP and K-means. It is mainly because that our algorithm takes the uncertainty into account, when using the term to express the topic of the data and gives the more important term a higher probability to indicate the topic of the data. It makes the data topic be expressed adequately. Another reason is that when calculate the probability that two pieces of data have the same topic, we consider

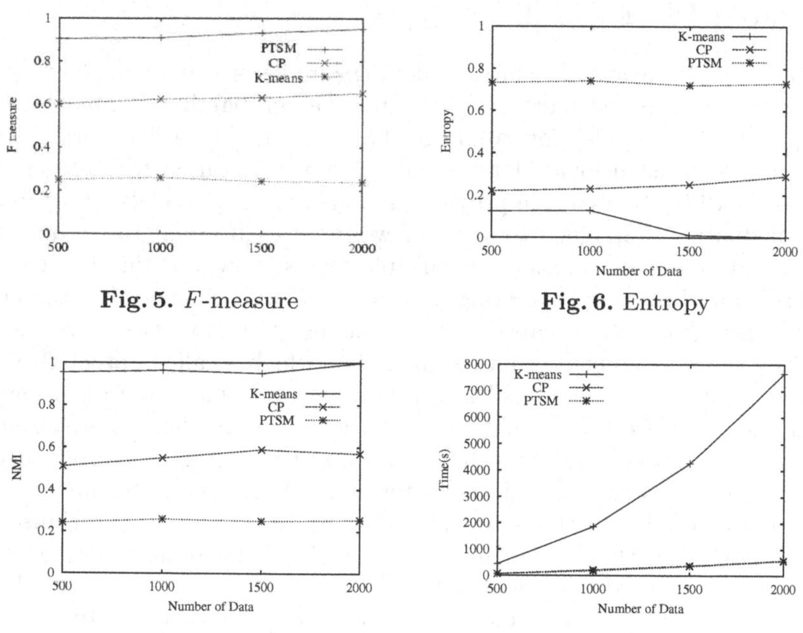

Fig. 5. F-measure

Fig. 6. Entropy

Fig. 7. NMI

Fig. 8. Execution time

not only the direct connection between the two pieces of data, but also indirect connections of them. It makes the probability be calculated more accurately. In CP algorithm, it takes the uncertainty of the document and the term into account. Both the probability of a word belonging to a word cluster and the probability of a document belonging to a document cluster are considered, but it does not think about the indirect relationships between data. K-means only compares the two pieces of data in the term-space and does not consider the semantic information between them. And also K-means is a kind of iterative method, only being local optimism. Thus, the clustering performance of K-means is the worst.

4.2 Model Efficiency

In this experiment, we compare our algorithm with CP and K-means algorithms in execution time. We test the execution time under data sets of different size. Figure 8 shows the comparing results of PTSM, CP and K-means.

The execution time of our algorithm is much less than that of K-means', but approximately equals to that of CP. For K-means, it is a iterative method, which always consumes a large amount of time. On the other hand, our algorithm filters most of the unnecessary terms in data clustering. In fact, it is a kind of feature reduction processing. The two above reasons make our algorithm's execution time be much less than K-means. Both CP algorithm and our model use matrix as the means of clustering, so the complexity of the two methods is similar and the execution time are approximately the same.

5 Related Work

In order to establish an automated indexing for discrete data in data space, Hofmann [9] has proposed another algorithm, called probabilistic latent semantic indexing (PLSI), which is motivated and based upon the earlier work. In this algorithm, a low-dimensional "latent semantic space" is used to represent the topic of data. PLSI, treated as a projection model, directly models latent topics, since probabilities is assigned to a set of words in each latent topic, and thus data can be treated as a mixture of multiple topics. However, this PLSI model is not built for clustering processing and is not a proper generative model, as pointed by Blei et al. [10], because it treats data as random variables and cannot generalize to new documents. Another algorithm, which is called Latent Dirichlet allocation (LDA) [10], generalizes PLSI by treating the topic mixture parameters as variables drawn from a Dirichlet distribution. This model is a well-defined generative model and performs much better than PLSI, but its clustering effect is still missing. On the other side, data clustering has been intensively investigated and the most popular methods are probably partition-based algorithms, just like K-means. Recently, focuses in document clustering domain shifted from traditional vector based document similarity for clustering to suffix tree based document similarity, as it offers more semantic representation of the text present in the document [11]. But uncertainty between terms and topic is not considered.

Non-negative matrix factorization (NMF) [12] is another candidate and is also shown to obtain good results in [13]. Tao Li [4] provide a mechanism which uses a nonnegative matrix factorization model $X = FSG^T$, where X is the word document semantic matrix, F is the posterior probability of a word belonging to a word cluster and represents knowledge in the word space, G is the posterior probability of a document belonging to a document cluster and represents knowledge in the document space, and S is a scaled matrix factor which provides a condensed view of X. Another closely related work is the so-called two-sided clustering, like [14] and [15], which aims to clustering words and data simultaneously. In [14], it is implicitly assumed a one-to-one correspondence between the two sides of clusters. [15] is a probabilistic model for discrete data, but it has similar problems as in PLSI and is not generalizable to new documents. All the work is concerned about data clustering, while lacking the probabilistic interpretations to the connections among data and take a comprehensive relationships between two data into account.

Probabilistic topic models were originally developed and utilised for document modeling and topic extraction in Information Retrieval. A brief overview of the latent semantic analysis techniques, i.e., Latent Semantic Analysis, probabilistic Latent Semantic Analysis, and Latent Dirichlet Allocation, is provided in paper [16]. In probabilistic topic models, only direct similar relationship among data is considered. However, PTSM can deal with up to k-connection similar relationships among data.

6 Conclusions and Future Work

In this paper, we propose a novel model based on probability and term similarity, called PTSM for data clustering. The most important term in each data for clustering is assigned a probability to indicate the topic of the data. And a probabilistic similarity model is established based on the different data expression. PTSM considers the uncertainty and a comprehensive consideration of the similarity between two pieces of data in clustering processing. We compared the results of our algorithms with improved k-means algorithm on a standard public datasets. And it outperforms both on effectiveness and efficiency. We simply assume that any two data are independent, which means they have no relationships. The future work towards PTSM includes considering the relationships of any two terms in the dictionary, commonly existing in real-life. Some papers propose to integrate multi-agent with data mining technologies [17–19], we are trying to use the idea to improve PTSM's performance further.

References

1. Li, G., Ooi, B.C., Feng, J., Wang, J., Zhou, L.: EASE: An effective 3-in-1 keyword search method for unstructured, semi-structured and structured Data. In: Proceedings of Special Interest Group on Management of Data, pp. 903–914 (2008)

2. Zobel, J., Moffat, A.: Inverted files for text search engines. ACM Computing Surveys 38(2), Ariticle 6 (2006)
3. Steinbach, M., Karypis, G., Kumar, V.: A comparison of document clustering techniques. Technical Report. University of Minnesota-Computer Science and Engineering, Minnesota (2000)
4. Li, T., Ding, C., Zhang, Y., Shao, B.: Knowledge transformation from word space to document space. In: Proceedings of Special Interest Group on Information Retrieval, pp. 187–194 (2008)
5. Larsen, B., Aone, C.: Fast and effective text mining using linear-time document clustering. In: Proceedings of Special Interest Group on Knowledge Discovery and Data Mining, pp. 16–22 (1999)
6. Van Rijsbergen, C.J.: Information Retrieval. Butterworth-Heinemann Ltd. (1989)
7. Kowalski, G.: Information retrieval systems: theory and implementation. Springer, 10.1016/S0898-1221(97)80229-5 (1998)
8. Strehl, A., Ghosh, J.: Cluster ensembles: a knowledge reuse framework for combining multiple partitions. The Journal of Machine Learning Research 3, 583–617 (2003)
9. Hofmann, T.: Probabilistic latent semantic indexing. In: Proceedings of Special Interest Group on Information Retrieval, pp. 50–57 (1999)
10. Blei, D.M., Ng, A.Y., Jordan, M.I.: Latent dirichlet allocation. Journal of Machine Learning Research 3, 993–1022 (2003)
11. Rafi, M., Maujood, M., Fazal, M.M., Ali, S.M.: A comparison of two suffix tree-based document clustering algorithms. CoRR abs/1112.6222 (2011)
12. Lee, D.D., Seung, H.S.: Learning the parts of objects with nonnegative matrix factorization. Nature 401, 788–791 (1999)
13. Xu, W., Liu, X., Gong, Y.: Document clustering based on non-negative matrix factorization. In: Proceedings of Special Interest Group on Information Retrieval, pp. 267–273 (2003)
14. Dhillon, I.S.: Co-clustering documents and words using bipartite spectral graph partitioning. In: Proceedings of Special Interest Group on Knowledge Discovery and Data Mining, pp. 269–274 (2001)
15. Hofmann, T., Puzicha, J.: Statistical models for co-occurrence data. Technical Report AIM, 1625 (1998)
16. Wang, W., Barnaghi, P., Bargiela, A.: Probabilistic Topic Models for Learning Terminological Ontologies. IEEE Transactions on Knowledge and Data Engineering, 1028–1040 (2010)
17. Cao, L.: Data Mining and Multi-agent Integration (edited). Springer (2009)
18. Cao, L., Weiss, G., Yu, P.S.: A Brief Introduction to Agent Mining. Journal of Autonomous Agents and Multi-Agent Systems 25, 419–424 (2012)
19. Cao, L., Gorodetsky, V., Mitkas, P.A.: A Agent Mining: The Synergy of Agents and Data Mining. IEEE Intelligent Systems 24(3), 64–72 (2009)

Following Human Mobility Using Tweets

Mahdi Azmandian, Karan Singh, Ben Gelsey,
Yu-Han Chang, and Rajiv Maheswaran

Information Sciences Institute
University of Southern California
Marina del Rey, CA 90292
{azmandia,karans,gelsey}@usc.edu,
{ychang,maheswar}@isi.edu

Abstract. The availability of location-based agent data is growing rapidly, enabling new research into the behavior patterns of such agents in space and time. Previously, such analysis was limited to either small experiments with GPS-equipped agents, or proprietary datasets of human cell phone users that cannot be disseminated across the academic community for followup studies. In this paper, we study the movement patterns of Twitter users in London, Los Angeles, and Tokyo. We cluster these agents by their movement patterns across space and time. We also show that it is possible to infer part of the underlying transportation network from Tweets alone, and uncover interesting differences between the behaviors exhibited by users across these three cities.

1 Introduction

Location-based agents are becoming increasingly prevalent, and the data generated by these agents is a rich domain for data mining and interaction research. This involves an important issue, i.e. mining agent data to enhance agent performance, an important topic in agent mining [1,2]. Agents are sometimes location-based advertising bots, location-based game virtual characters, or humans using GPS-capable devices such as smartphones. Understanding location-based behavior can lead to better models of people and cities and help improve decision-making in domains from transportation networks to advertising. In this paper, we focus on geotagged data generated by Twitter-users, and apply data mining and visualization techniques to uncover both behavior patterns as well as the underlying network structure that supports the agent movements in London, Los Angeles and Tokyo. Understanding location-based behavior can be used to build more accurate models of human movement, which could then be deployed to any number of applications ranging from transportation modeling to personalized and predictive location-aware agent services to assessing "patterns of life" in foreign cities and towns.

We first introduce the notion of a *trace*, which is simply a user's trajectory extracted by connecting his tweet locations through the course of a day. Traces are broken down to fragments that correspond to periods where a user is tweeting

L. Cao et al.: ADMI 2012, LNAI 7607, pp. 139–149, 2013.

frequently. These fragments will also include updates of the user's location, which yields relatively accurate knowledge of the user's location during such a fragment. These fragments are used to construct a visualization we call a Trace-Based Heatmap. We then demonstrate an algorithm that can infer an undirected graph depicting the routes in the city where tweeting is most active. We also apply clustering techniques to this spatio-temporal data, and show that users can be roughly described by their geographic area and temporal description of their Twitter use. These results show initial promise towards agent models that can be learned from publicly available geo-tagged data sources.

2 Related Work

With the growing prevalence of social media such as Facebook and Twitter, researchers in social and network science have shown great interest in the datasets generated. Recently GPS-tagged information and "check-ins" have become quite widespread, giving rise to a new field of location data analysis [3]. In the past, the majority of research used human location data procured through mobile phone networks. These studies range from behavioral predictions [4], development of human movement [5], detecting anomalies [6], identification of points of interest from trajectories [7], discovery of the most popular routes [8], trajectory clustering [9], identification of movement flocks [10], and inference of transportation routines [11]. Such extensive research is justified considering the broadness of the potential applications, running the gamut from urban planners on the search for discovering daily routines [12], to biologists modeling the worldwide spread of pandemic influenza [13]. Similar techniques have also been used in ecology to track animal movements [14].

Visualization is a crucial tool in this human mobility research. Visualizations enhance tangibility of data mining outcomes and guide computational methods, providing a compensation for the computer's inability to incorporate humans' tacit knowledge [15]. Mobility data visualization has come a long way from the elementary idea of drawing arrows on an image [16] simply indicating direction of movement. Time-Geography study introduced the "space-time cube" technique; approaches to managing large-scale data have suggested data aggregation techniques such as the temporal histogram, traffic density surface, and accessibility surface; data filtering according to user-specified queries has been an alternative approach to handling large amounts of data [17]; and in a more recent endeavor a multidisciplinary approach was applied to develop a framework for the analysis of massive movement data taking advantage of a synergy of computational, database, and visual techniques [15].

Among these visualization techniques, the following approaches are most relevant to our work: The first approach is based on spatial, temporal or attribute proximity [18] (the space, time or attribute space are divided into compartments, into which the trajectories (viewed as a set of discrete movement events, i.e. geographic locations with respective time stamps) are projected). The second approach is also trajectory-based where trajectories are aggregated in their entirety based on their similarity in geographic, temporal or attribute space (or a combination thereof) [19]. The "route-based" aggregation [18] is often performed by clustering in the data space or in an abstract projection thereof [20]. The third approach aggregates movement data based on their origin and destination and ignores the route between these two spatial locations, so that the movement is seen as a vector between the two locations, not as a set of recorded positions on a trajectory [18].

The novelty of our approach is that the data we use comes from "geotagged" tweets where we create trajectories via notions of "traces" and "fragments". We then develop various algorithms to turn these into heatmaps, route graphs, flows, behavior characterizations and temporal signatures for cities.

3 Data Visualization

Heatmap Constructions. Two types of heatmaps were generated for each area of study, a "Point-Based HeatMap" and a "Trace-Based HeatMap", the description of which will follow. For the Point-Based HeatMap, for every single tweet occurring at a particular gridpoint, an intensity increment of 3 units was applied to the cell, 2 units of intensity to the surrounding 8 cells, and one unit of intensity to the 16 cells encompassing the previous 9. For the Trace-Based HeatMap, for each fragment in each trace of each user the following was done: Each line in a fragment, was mapped to a discretized line on the grid using "Bresenham's Line Algorithm" [21]. Each line on the grid, contributed to two units of intensity incrementation on each point residing on it; also for the two parallel adjacent discretized lines to the previous, one additional unit of heat was introduced on each point residing on them. We name this method "Radial Line Heat Application".

Route Graph Extraction. Given the trace-based heat map as input, we introduce an algorithm to extract the underlying transportation network upon which the Twitter users are moving. The algorithm proceeds in a greedy manner, identifying potential edges which contain the highest local intensity of traversal by Tweet traces. Intuitively, this corresponds to the lines of red in Figure 1 on page 146. These identified edges are initially short, and through an iterative procedure, they are extended along directions with high Tweet traversal. A few additional tricks are needed to prevent an excess of edges being identified in regions where there is intense Tweeting spread out over a wide area, such as preventing the discovery of new additional edges that are nearly identical to previously identified edges. The pseudo-code is provided in Algorithm 1.

The algorithm keeps track of areas that it has already searched by setting map cells as being "engaged" once an edge has been found nearby. Initially, all cells are set as "disengaged". The algorithm then follows an iterative procedure in which, for every iteration, the following procedure is executed: A grid cell is chosen which has the highest amount of heat among all the "disengaged cells", and a new vertex defined on that location is added to the graph. Every disengaged grid cell within a radius of $searchRadius = 15$ cells is considered as a candidate for the next vertex to add to the graph, along with the edge that connects the two. Each of these candidates are scored by summing the amount of Tweet traversal intensity along the edge. The grid cell candidate with the highest score determines the location of next vertex and is added to the graph. The edge connecting the previous two vertices is also added to the graph.

Next, the algorithm attempts to extend the new edge in both directions. On each direction, similar to before, all disengaged grid cells within a radius of $searchRadius = 15$ cells are considered as candidates for the next vertex to add to the graph, but this time, vertices that would result in an edge extension with an angular deviation of more than 15 degrees are disregarded. This restriction is intended to ensure that the path being formed corresponds to a single road on the map. If the highest scored vertex has an average heat of more than $thresholdRatio = 0.8$ times the average Tweet traversal intensity of the edge to be extended, it will be added to the graph along with its corresponding vertex, otherwise extension in this direction will reach cessation. During each attempt of extension, if an existing vertex is found within the search radius, and this vertex has an edge which forms an angle of less than 15 degrees with the edge to be extended, the two edges are connected and the process stops. This also prevents having redundant edges denoting essentially the same path. After path extension in both directions is complete, all grid cells within a radius of $engagingRadius = 10$ from the new edge is flagged as "engaged".

Patterns of Life. In the previous section, we use the aggregated data of all the users' activity traces to infer the underlying transportation network which guided the trajectories of the users. Here, we demonstrate a simple clustering technique to understand the different classes of user behavior. First, we apply K-means clustering on the dataset containing all the coordinates of each Tweet in our dataset. This results in clusters representing broad geographic areas where Tweeting activity occurs. We use Dunn indexing to choose an appropriate K. Given these geographic regions, we then create an activity vector v for each user:

$$v = [v_1^1 v_1^2 .. v_1^K v_2^1 v_2^2 .. v_2^K .. v_7^1 v_7^2 v_7^K],$$

where v_j^i is the Tweeting activity level for this user on the jth day of the week in geographic region i. This activity vector is normalized so that Tweeting activity sums to one for each day of the week. We then apply a second K-means clustering to this new set of vectors.

4 Data

Our dataset is extracted from Twitter, a popular micro-blogging service. In the Twitter terminology, the microblog messages or "tweets", are equipped with the option of containing what is referred to as "geotags". Geotags are labels that indicate where a Twitter user was, when the tweet was posted. On a client's side, a geotag can be applied by activating the geotag functionality in the settings of the twitter application being utilized.

Algorithm 1. Graph Route Extraction given *heatmapGrid*

```
{heatmapGrid is assumed to store heat values assigned to each grid cell}
{Graph G is initially empty and gridCellFlags is a grid of booleans initially set to
false .}
{The distance between two gridpoints is the number of cells along the line connecting
them with Bresenham's algorithm, which is equal to their Chebyshev distance}
searchRadius ← 15
engagingRadius ← 10
thresholdRatio ← 0.8
for i = 1 → numberOfIterations do
  gridPoint0 ← gridpoint with most heat value among disengaged gridpoints
  Add a new vertex v0 to G with location defined as gridPoint0's location
  for every gridPoint within a radius of searchRadius from gridPoint0 do
    sum ← 0
    for every midGridPoint that appears along the line connecting gridPoint and
    gridPoint0 do
      {The line connecting two grid points is determined with Bresenham's Line
      Algorithm}
      sum ← sum + heatmapGrid[XOf(midGridPoint)][YOf(midGridPoint)]
    Assign sum as the score for gridPoint
  pathEdges ← {}
  gridPoint1 ← gridPoint with the highest score
  Add a new vertex v1 to G with location defined as gridPoint1's location
  Add a new edge e = {v0, v1} to G
  addetopathEdges
  directions ← {(gridPoint0, gridPoint1), (gridPoint1, gridPoint0)}
  while !isEmptydirectionsdirection do
    direction = /textremoveFirstElement(directions)
    extendInDirection(direction)
  for every edge e in pathEdges do
    for every midGridPoint that appears along e do
      for every gridPoint within a radius of engagingRadius from midGridPoint
      do
        gridCellFlags[XOf(gridPoint)][YOf(gridPoint)] ← true
```

A tweet logged would contain exact latitude and longitude coordinates if and only if the tweet was sent through a smartphone (or any hand-held GPS-equipped device with the geotagging functionality switched on); Otherwise a tweet will include more general geotags like "Santa Monica" or "Marina Del Rey"; or perhaps lack any type of geotag whatsoever. One of the services Twitter's Streaming API provides, is live-streaming all tweets originating from a predetermined coordinate range (known as the "filter" service). This implies all such returned tweets will have non-empty geotag fields. Among retrieved tweets, those without latitude and longitude coordinates (roughly half of them) were discarded. In this paper, we focus on London, Los Angeles and Tokyo. From September 18, 2011 to February 9, 2012, we collected 22,496,299 tweets geotagged with latitude and longitude information.

Algorithm 2. Path Extension given *direction*

```
head = directionHead(direction)
tail = directionTail(direction)
threshold = thresholdRatio × averagePathHeat(head, tail)
for every gridPoint within a radius of searchRadius from head do
  angleDeviation = LineAngleDifference(line(head, tail), line(head, gridPoint))
  if angleDeviation/leqπ/12 then
    connectToVertex ← false
    verticesFound ← {}
    if hasVertex(gridPoint) then
      for every edge link originating from gridPoint do
        angleDeviation = LineAngleDifference(line(head, tail), link)
        if angleDeviation/leqπ/12 then
          connectToVertex ← false
          add gridPoint to verticesFound
    if isEmpty(verticesFound) then
      sum ← 0
      cellCount ← 0
      for every midGridPoint that appears along the line connecting gridPoint
      and head do
        sum ← sum + heatmapGrid[XOf(midGridPoint)][YOf(midGridPoint)]
        cellCount ← cellCount + 1
      Assign sum/cellCount as the average score for gridPoint
if !isEmpty(verticesFound) then
  v_con ← nearest vertex to head in verticesFound
  Add a new edge e = {v_con, head} to G
  addetopathEdges
else
  gridPoint_con ← gridPoint with the highest score
  if score(gridPoint_con) ≥ threshold then
    Add a new vertex v_con to G with location defined as gridPoint_con's location
    Add a new edge e = {v_con, head} to G
    addetopathEdges
    add direction {head, v_con} to directions
```

Data Processing. Among the many users tweeting via their phones, the most useful are those who tweet frequently throughout the day. By having the tweets for a such user, one can attempt to extrapolate the person's location throughout the day. We define a *trace* to be all tweets for one user starting at 4:00 AM of a day and ending at 3:59 AM of the following day. Empirically, it was the time of lowest activity in all the cities we were analyzing. In order to gauge the value of a particular trace of a user's daily activity, we use a "Pulsating Heuristic". Each time a person tweets it "pulsates", i.e., a radius of 400 meters (which is roughly the GPS error in our tweeting data which was calculated empirically) is "affected" by a pulse sent out from the tweet location. The lifespan of this pulse effect is set to half an hour. Once a person tweets again, the effect of previous pulses will be nullified. Now for evaluating a tweet and assigning a value to its utility, starting from a score of zero, each time a person tweets from a non-effected location, the score is incremented. This heuristic avoids giving weight to users that frequently tweet but always at the same location. Instead it gives more weight to users that frequently tweet from different locations, even if the number of distinct locations is small.

Bot Filtering. Many geotagged tweets in our dataset were sent by bots sending location-specific news or advertisements. Although one might assume bots to be stationary (resulting in all tweets originating from a fixed location), certain services have multiple stations which all send information through the same alias. Traffic and incident reporting services and dining-venue-advertising functions would best exemplify such users. We use two heuristics to filter out most of these messages: (1) if a user is tweeting URLs (identified by finding the substring "http" in the tweet) more than 70% of the time, it is likely to be a bot, and (2) if a user's movement speed is more than 120 km/h, it is likely to be a bot (speed being calculated as the straight line distance between two consecutive tweeting events).

Trace Fragmentation. We next process the data by segmenting each day's trace into one or more fragments that represent a period of continuous user movement. For every trace, after displaying a point on the map representing the location of a tweet occurrence, the assigned "trajectory" to each trace was defined as the jagged line created by connecting each tweet's corresponding point, to the point of its next tweet. The idea was to extrapolate one's daily trajectory. One issue is that drawing a direct line between two points is too poor of an estimation, unless the two occurrences are fairly close in time. Therefore given a trace, we define a notion of a "fragment" as a maximal sub-path in a trajectory which for every two consecutive tweet occurrences, the following two conditions hold: (1) the time between the two is less than 30 minutes, and (2) the location distance is more than 400 meters apart. The former condition is enforced to reduce extrapolation error, and the latter to avoid scenarios in which the observed displacement is merely a result of GPS error. The notion of trajectory fragments is the basis of the work in this paper.

Location and Area Discretization. In order to facilitate the process of HeatMap construction, the area of study was sliced into a grid where each cell covers a 0.0005 degrees of latitude by 0.0005 degrees of longitude surface (being roughly equivalent to a 50 by 50 meter coverage). After executing necessary heat application calculations (stated earlier), for visualization purposes, this grid was in turn "flattened" using the Mercator method.

5 Results

Transport Network Inference. When comparing cities it seems that people in Tokyo make much more use of the public transportation, and this is characterized by seeing heat on particular routes in the Trace HeatMap that have blobs of red in the Point HeatMap for Tokyo. Such paths were actually bus or subway routes and the blobs of red corresponded to the location of stops or stations; and example of which is shown on Figure 3.

By comparing the visualization results of our "Point-Based HeatMap" and the "Trace-Based HeatMap" displayed in Figure 1, clearly the underlying transportation routes are significantly more visible in the latter. Heat is sometimes seen to be relatively higher on portions of routes in the Point heatmap due to the fact that naturally, the more time people spend in an area, the more likely it is to have tweet activity; of course time spent in routes for each individual may not be high, but in a large scale, routes are densely populated.

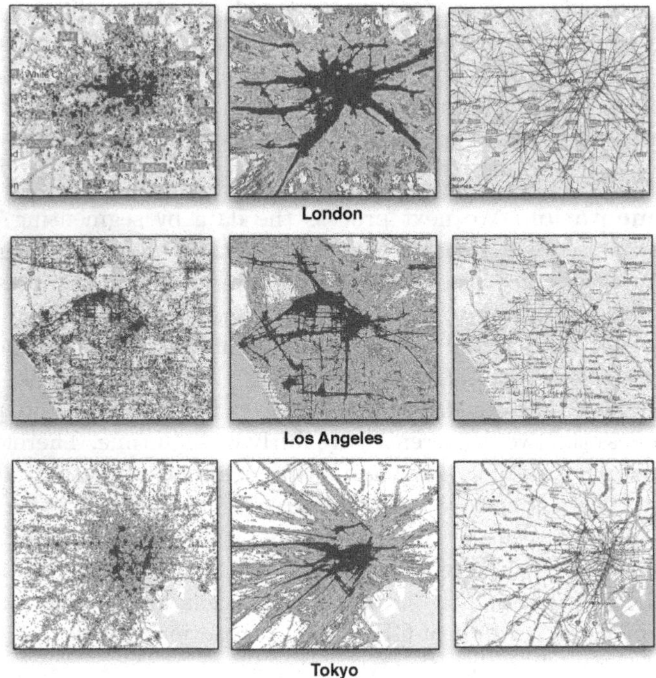

Fig. 1. Point-Based HeatMap, Trace-Based HeatMap and Route Graph of London, Los Angeles and Tokyo

We also managed to identify area-specific phenomena through studying the visualizations. Many of the main routes are nicely highlighted, as shown in the Figure 2, the 405 freeway is virtually inactive; which shows that tweeting on the 405 is not a common practice.

For the graph construction algorithm, an example of its functionality can be seen in Figure 4 which parts of Sunset Boulevard and West Hollywood in Los Angeles are identified.

Patterns of Weekly Activity. No surprises were uncovered in our analysis of weekly activity patterns. Our algorithm reported nine broad geographic areas of

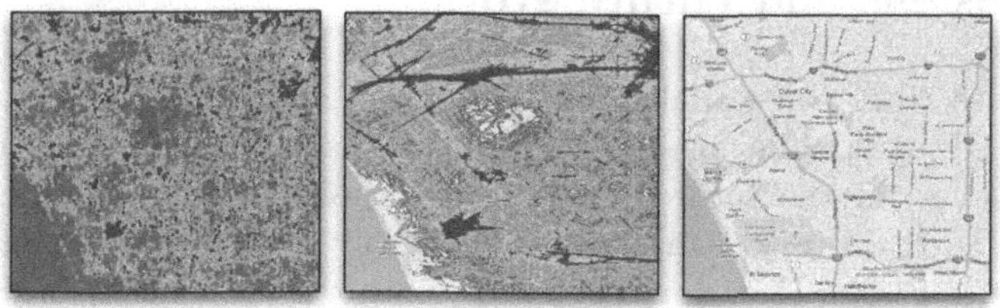

Fig. 2. Point-Based HeatMap, Trace-Based HeatMap and Map of Los Angeles Showing the 405 Freeway Not Properly Delineated in Visualization

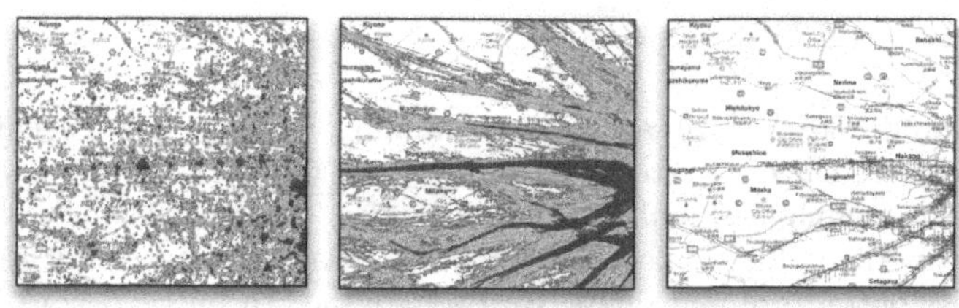

Fig. 3. Point-Based HeatMap, Trace-Based HeatMap and Route Graph of Tokyo, Uncovering the Existence of a Public Transportation Route

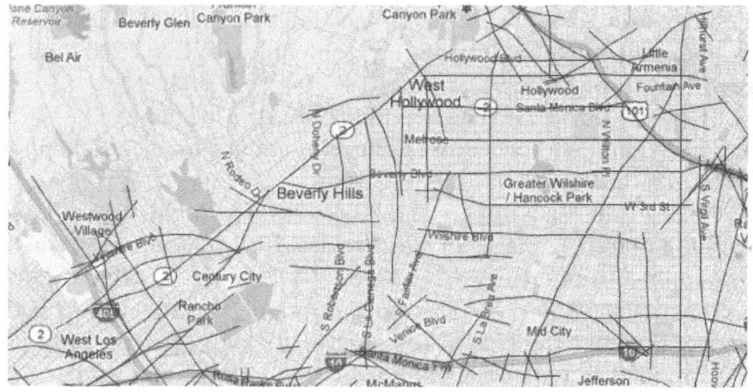

Fig. 4. An example of the Graph Route Construction Algorithm's Efficacy

tweeting activity within the Los Angeles dataset. User behavior was clustered into eight classes, with each of these classes roughly corresponding to users whose activity was mainly centered in one of the broad geographic areas. The results we got by applying our Algorithm (to find pareiodic patterns in movement of people based on K reference points) to our Los Angeles Data set were not surprising. There was some differentiation between activity on weekdays and weekends, with users exhibited slightly higher entropy in terms of their distribution of activity over the nine geographic regions. This makes intuitive sense, since people tend to travel more for leisure on weekends.

6 Conclusions and Future Work

This paper describes the visualization, analysis, and mining of location-based data generated by mobile agents. Here, our analysis focused on geo-tagged Tweets. However, in the future, we hope to apply these techniques to other agent-generated trajectory data, such as agents in location-based games, UAVs, or indeed Twitter bots. Discovery of spatio- temporal patterns in such data is an important and challenging problem, and here we have only presented an initial step in this direction.

References

1. Cao, L., Gorodetsky, V., Mitkas, P.: Agent mining: The synergy of agents and data mining. IEEE Intelligent Systems 24(3), 64–72 (2009)
2. Cao, L.: Data mining and multi-agent integration. Springer, Dordrecht (2009)
3. Noulas, A., Scellato, S., Mascolo, C., Pontil, M.: An empirical study of geographic user activity patterns in foursquare. In: Proc. of the 5th Int'l AAAI Conference on Weblogs and Social Media, pp. 570–573 (2011)
4. Song, C., Qu, Z., Blumm, N., Barabási, A.L.: Limits of predictability in human mobility. Science 327(5968), 1018–1021 (2010)
5. Azevedo, T.S., Bezerra, R.L., Campos, C.A.V., de Moraes, L.F.M.: An analysis of human mobility using real traces. In: Proceedings of the 2009 IEEE Conference on Wireless Communications & Networking Conference, WCNC 2009, pp. 2390–2395. IEEE Press, Piscataway (2009)
6. Candia, J., Gonzalez, M.C., Wang, P., Schoenharl, T., Madey, G., Barabasi, A.L.: Uncovering individual and collective human dynamics from mobile phone records. Math. Theor. 41, 224015 (2008)
7. Palma, A.T., Bogorny, V., Kuijpers, B., Alvares, L.O.: A clustering-based approach for discovering interesting places in trajectories. In: Proceedings of the, ACM Symposium on Applied Computing, SAC 2008, 863–868. ACM, New York (2008)
8. Chen, Z., Shen, H.T., Zhou, X.: Discovering popular routes from trajectories. In: Proceedings of the 2011 IEEE 27th International Conference on Data Engineering, ICDE 2011, pp. 900–911. IEEE Computer Society, Washington, DC (2011)
9. Masciari, E.: A Framework for Trajectory Clustering. In: Trigoni, N., Markham, A., Nawaz, S. (eds.) GSN 2009. LNCS, vol. 5659, pp. 102–111. Springer, Heidelberg (2009)
10. Vieira, M.R., Bakalov, P., Tsotras, V.J.: On-line discovery of flock patterns in spatio-temporal data. In: Proceedings of the 17th ACM SIGSPATIAL International Conference on Advances in Geographic Information Systems, GIS 2009, pp. 286–295. ACM, New York (2009)
11. Liao, L., Patterson, D.J., Fox, D., Kautz, H.: Learning and inferring transportation routines. Artif. Intell. 171(5-6), 311–331 (2007)
12. Sevtsuk, A., Ratti, C.: Does urban mobility have a daily routine? learning from the aggregate data of mobile networks. Journal of Urban Technology 17(1), 41–60 (2010)
13. Colizza, V., Barrat, A., Barthelemy, M., Valleron, A.J., Vespignani, A.: Modeling the worldwide spread of pandemic influenza: Baseline case and containment interventions. PLOS Med. 4, e13 (2007)

14. Li, Z., Ji, M., Lee, J.G., Tang, L.A., Yu, Y., Han, J., Kays, R.: Movemine: mining moving object databases (2010)
15. Andrienko, G., Andrienko, N., Wrobel, S.: Visual analytics tools for analysis of movement data. SIGKDD Explor. Newsl. 9(2), 38–46 (2007)
16. Vasiliev, I.R.: Mapping Time. Cartographica 34(2) (1997)
17. Kapler, T., Wright, W.: Geo time information visualization. Information Visualization 4(2), 136–146 (2005)
18. Gennady, A., Natalia, A.: A general framework for using aggregation in visual exploration of movement data. Cartographic Journal 47(1), 22–40 (2010)
19. Laube, P., Imfeld, S., Weibel, R.: Discovering relative motion patterns in groups of moving point objects. International Journal of Geographical Information Science 19, 639–668 (2005)
20. Skupin, A., Hagelman, R.: Visualizing demographic trajectories with self-organizing maps. Geoinformatica 9(2), 159–179 (2005)
21. Bresenham, J.E.: Algorithm for computer control of a digital plotter, pp. 1–6. ACM, New York (1998)

Part IV

Agent Mining Applications

Agents and Distributed Data Mining in Smart Space: Challenges and Perspectives

Vladimir Gorodetsky

SPIIRAS, 39, 14-th Liniya, St. Petersburg, 199178, Russia
gor@iias.spb.su

Abstract. Smart space is a distributed ambient environment with existing, inside it, dynamic set of inhabitants (living and nonliving) solving various own and common tasks. The mission of smart space is to provide, for its inhabitants, with context–dependent information, communication, services, reminders and personalized recommendations in a user–friendly mode where and when needed in ubiquitous and unobtrusive style. The smart space R&D uses large diversity of models, frameworks, and technologies and their integration is the first challenging smart space problem. Another challenge is caused by the necessity to process huge volumes of heterogeneous information perceived by distributed sensors in adaptive, self–organizing, learnable, and efficient style. The paper analyses these challenges and emphasizes an important role of the technology integrating agent and data mining to overcome both these challenges.

1 Introduction

Smart space is a distributed ambient environment with existing, inside it, dynamic multitude of inhabitants (living and nonliving, e.g. robots, entities) solving their own and common tasks. The mission of this environment is to provide, for smart space inhabitants, with context–dependent information, ubiquitous communication, personalized services, reminders and personalized recommendations in a user–friendly mode where and when needed in ubiquitous and unobtrusive style. This definition covers many important modern applications, e.g. smart home, smart city, spatial security systems, emergency management, environmental monitoring, health care and disability person's assistance, smart car cabin, smart grid, and many others.

This area is currently paid a lot of attention, in particular, within FP-6 and FP-7 programs of European Commission ([6], [16], [17], [19]). E.g., *Angel* project [1] (2006–2009) was devoted to the development of a platform supporting smart *wearable body sensor network* intended to monitor of a patient health data (temperature, pulse, breathing, etc.), his/her behavior (using microphones and cameras), as well as environmental climate. These sensors are used in two modes. The first one has to periodically inform the medical staff on the patient's health state, whereas the second aims to detect situations that are dangerous either for the patient's health, or for the environment as a whole. The examples of the latter are situations requiring immediate intensive medical cares, alarming about

L. Cao et al.: ADMI 2012, LNAI 7607, pp. 153–165, 2013.

the fire to fire–brigade, police and to the medical staff, etc. Other examples are the projects AWARE [2], Hydra [12], MORE [15], SMEPP [18], etc.

As a rule, smart space systems are composed of heterogeneous ad-hoc wireless sensor network, effectors, as well as of diverse software intended for monitoring and on-line processing data produced by the sensors, performing intelligent data analysis, human's behavior pattern recognition to predict his/her needs, classification, context detection, user profile learning, etc. It uses multi-modal user interface reacting on spoken or gestured user activity including even an intelligent dialogue. Thus, smart space software is designed using multiple frameworks, models, and technologies requiring integration and adaptation of many smart space subsystems to the particular user needs, learning-based re-configuration of networked computation, communication, sensing and action that determine the first challenging problem. The second challenge is caused by the necessity to efficiently process huge amount of heterogeneous data and information perceived by distributed sensors as well as knowledge in order to provide, for the smart space, with intelligent, adaptive, self-organizing, learnable, actionable and efficient performance.

The paper objective is to highlight and justify a remarkable contribution that can be provided by integrated use of agent and data mining for managing of the aforesaid challenges. It shows that such integrated technology, even in its current state, can significantly contribute to smart space field. Indeed, synergy of agent and data mining technologies [3,4,5] makes easier to effectively solve several important classes of tasks. Among them, the tasks like data and information fusion, on-line learning of users' profiles using smart space log data and footprints of user's interaction with the web. Other specific tasks are sensors network and distributed software self-configuration, multi-source data mining, distributed and peer-to-peer (p2p) data mining and many other difficult tasks peculiar to smart space that can be solved using agent and data mining integrated technology.

The rest of the paper is organized as follows. Section 2 describes the author's view on generic smart space architecture composed of large number of agents interacting in distributed p2p style. It highlights many specific data mining/knowledge discovery and processing tasks to be solved by the smart space agents using distributed p2p data mining technologies. Section 3 advocates an important role of the agent-based technology in solution of the network self-configuration problem on the basis of p2p distributed learning implementing, in fact, a self-organization model. Section 4 describes the role of multiple contexts and presents a framework for context-driven data mining. Conclusion summarizes the paper results and outlines perspectives.

2 Smart Space Generic Architecture

To the present days, generally adopted smart space architecture does not exist. A variant of such architecture, according to the author's view, is depicted in Fig. 1. It is designed according to the functional view while accounting such requirements as re-configurability and personalization of the software and hardware to the particular user needs, necessity, for software components, to be simply extendable, adaptability to new services, autonomous learning of the user's preferences, etc.

In this architecture, the information field is constituted by the agents consuming data perceived by heterogeneous sensors. These agents can be capable to (pre)processing the input data, interact with each other and send the resulting data and information to particular applications. Sensor agent interaction can be destined to jointly (pre)process own information and then send the result to a service, e.g. when a number of sensors share participation in human behavior tracking and prediction. Other example is distributed security when security violation footprints are detected on the sensor level. On the other hand, sensors possess the limited computational resources (processor speed, memory), and their interaction can be necessary to solve more complex or more sophisticated tasks locally, at the sensor level, than each of them could be capable to solve separately.

Other data, information and knowledge sources also exist (Fig. 1). In many cases smart space is provided with the access to web-based information and services. Knowledge-intensive sources of information are data and knowledge bases, ontology, behavior knowledge base storing some frequent behavior patterns of humans living inside the smart space. One of the most important components

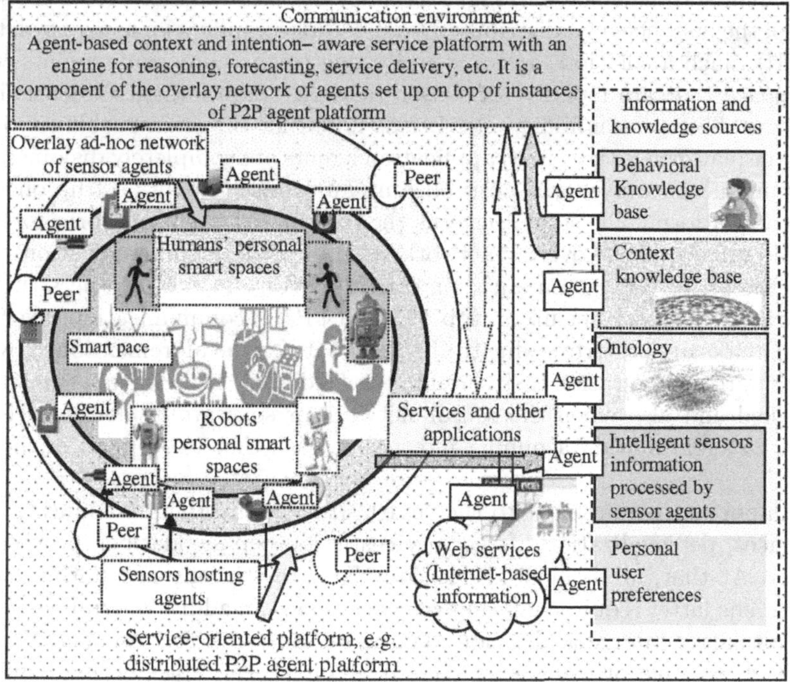

Fig. 1. Smart space architecture and software components. An example for smart home.

of the smart space information field is constituted by knowledge base storing multiple human preferences, because the focus of a smart space is to provide for human with personalized services and recommendations in predictive style. An important component of smart space architecture is so-called *service platform* [21] that is a context and intention– aware software component containing an engine destined for service delivery, context extraction, reasoning, forecasting, decision making, etc. In fact, the service platform should be an intelligent core of any smart space [21].

This architecture is agent-based. It is assumed that agents solve basic smart space tasks in distributed way, form various coalitions to efficiently solve particular tasks, while operating in p2p mode. They play the roles of mediators among many data, information and knowledge sources, take care about services personalization using smart space log data and footprints of the users' activity and behavior.

Agent interaction is supported by a software infrastructure known as *agent platform*. It is evidently that a conventional platform like JADE does not fit the smart space requirements due to existence of distributed nomadic components constituting the information field. E.g. home robots can be the carriers of the mobile component of smart space and they, due to limited distance of wireless communication channels, can dynamically leave and join the agent community of the stationary part of the smart space software and communication environment.

The architecture depicted in Fig. 1 assumes that smart space agents are set up on top of the instances of a distributed (e.g., FIPA-compliant) p2p agent platform ([9], [11]). Such a solution looks naturally due to the core requirements to the smart space that are mobility of sensor carriers, adaptability, self-learnability, re-configurability, extendibility, etc. Let us note that the main peculiarity of the p2p agent platform is that, in it, application agents are completely disjoined from the communication layer while establishing an overlay network set up on top of the agent platform. The same concerns the distributed agent platform instances that also constitute the overlay network set up on top of communications layer. Again, the set of p2p communication providers (Fig. 1) can also be structured as an overlay network set up on top of the TCP/IP communication layer. The latter remarks are important due to the fact that, for the aforementioned overlay network, a re-configuration task should be solved every time when a smart space is installed and also when the sensor network or software is modified. This is exactly an agent and data mining interaction task that is outlined in the next section.

An important requirement to the smart space operation is accounting user's preferences, personalization of the services and recommendations provided for the user. At that, all associated decisions have to strongly accounting current context. The latter requirement, in fact, determines an additional dimensionality to the all data mining and decision making tasks to be solved in smart space.

A good example of using agent-based architecture implementing a number of the aforementioned tasks is MAVHome architecture [7]. It is a pioneering work in accounting context in data mining/knowledge discovery and decision making. This architecture comprises of *cooperating layers*. Its bottom is constituted by perception layer composed of the sensors intended for the monitoring of the environment and providing, for smart space, with information via the interface layers. This information is also stored in the database and further used to extract inhabitants' behavior patterns through context–driven data mining. Learned decision makers (at the next layer) produce decisions and prediction while receiving new input information. These decisions are sent to applications (on the top layer) on request or according to a priory arranged strategy. Action execution is done in the top–down basis. The decision action is communicated to the service layer, which records the action and communicates it to the physical components. The physical layer, in its turn, performs the action using power line control and other automated hardware that is changing the state of the world model and triggering to new perception. Thus, MAVHome architecture solves several agent–related data mining tasks to support adaptive operation of smart space. In particular, it accumulates historical data, supports sensor data mining and higher-level data mining using reinforcement learning, implements adaptation and self–configuration as well as decision fusion using arbitration.

Other agent learning tasks are also peculiar for smart space. Among them, user preferences and profile mining, predictive service recommendation, context extraction, and some others are of the primary importancea. The subsequent sections describe some of these challenging tasks in more details and outline approaches that are implementable only on the basis of integration and interaction of the agent and data mining technologies.

3 Smart Space Self–configuration and Coalition Formation through Distributed Agent Learning

The basis of a smart space architecture depicted in Fig. 1 is using of p2p networking, in particular, overlay networking. This means that an important task is selection, by each agent of the network, its virtual neighbors. The choice of the virtual neighbors can be done in various modes using different criteria, etc. This task is well known as p2p network configuration. To say in more simple words, the p2p network configuration task can be formulated as the search for an answer on the question: "Who about whom has to know in the p2p network to solve the task in the best way?".

Each overlay network of a smart space can change its topology due to many causes. E.g. some nodes of a smart space network can leave the network or join it. The examples are mobile smart space of home robots, nomadic devices of the smart space inhabitants, etc. Wireless nomadic sensors can leave the network due to discharge of their batteries. Communication network can be supported by unmanned aerial vehicles if smart space is constituted to manage an emergency in an area where communication infrastructure is absent. New sensors and new software can appear in smart space and therefore the p2p networks constituting smart space cannot be finally configured from the very beginning. A specific of smart space networks is that the aforementioned network configuration tasks have to be solved by the smart space itself, without human intervention. Therefore this task is usually called self–configuration and it is one of the challenging tasks of the smart space domain. Let us consider general ideas concerning how it can be solved through integration of the agent and distributed data mining technologies.

Below the developed and experimentally validated generic (criteria–independent) network (re)configuration distributed algorithm (protocol) is outlined. In fact, it implements p2p learning algorithm [10]. For every particular case this generic protocol should be specialized while mostly taking into account specific optimization criterion used. Let us note that in [10] this algorithm was applied to the distributed learning of p2p intrusion detection while implementing on–line reconfiguration of p2p intrusion detection agent network. Let us outline the basic idea of this algorithm.

It is assumed that overlay p2p network of application agents is initially configured in a way, e.g. by experts or the network structure is assigned randomly. A peculiarity of the algorithm is that there is no server specifically dedicated to this task and each network node (corresponding, e.g., to an application agent, or to an instance of a p2p agent platform, or to a p2p communication provider) can interact only with its (virtual) neighborhood using a p2p protocol. Let us consider interactions of an arbitrary network node X_i which, at starting point, is assigned virtual neighborhood $\mathbf{N}_i = \{X_{i_1}, \ldots X_{i_k}\}$, where k is the cardinality of the node X_i neighborhood.

1. Select (e.g., at random), a subset of nodes $\Delta\mathbf{N}_i$ containing nodes other than those from the set $\mathbf{N}_i$.

2. Start normal system operation with the set of X_i neighbors $\bar{\mathbf{N}}_i = \mathbf{N}_i \cup \Delta\mathbf{N}_i$, while recording, at some discrete time instants $1, \ldots, k, \ldots$, the set of inputs $\{y_{i_1}^{(k)}, \ldots, y_{i_s}^{(k)}, y_{j_1}^{(k)}, \ldots, y_{j_r}^{(k)}\}$ produced by the X_i virtual neighbors $X \in \{\mathbf{N}_i \cup \Delta\mathbf{N}_i\}$ of the sets $\mathbf{N}_i = \{X_{i_1}, \ldots X_{i_r}\}$ and $\Delta\mathbf{N}_i = \{X_{j_1}, \ldots X_{j_s}\}$, and set of values $\{Q_{i_1}^{(k)}, \ldots Q_{i_s}^{(k)}, Q_{j_1}^{(k)}, \ldots, Q_{j_r}^{(k)}\}$ evaluating a quality of the neighbors outputs. E.g., let a p2p agent system solves an intrusion detection task in distributed mode, and y_j is a reaction of the agent A_j set up on the node X_j, that is a neighbor of the X_i, in a situation when the agent A_i of the node X_i produces the signal *Alert*. In this case, each record $Y_i^{(k)} = \{y_{i_1}, \ldots y_{i_k}, y_{j_1}, \ldots, y_{j_s}\}$ corresponds to the reactions of the agent A_i neighbors when the latter produces the signal *Alert*. Let, afterwards, each such record is assigned, by the security administrator, a class label from the set $\{normal, attack\}$.

3. After n steps, the set of the interpreted records is accumulated by the agent A_i and the latter can use this sample as learning data to evaluate correctness of reactions (signals) of each its neighbor A_j. When evaluation is computed, agent A_i selects, from the set $\bar{\mathbf{N}}_i = \mathbf{N}_i \cup \Delta\mathbf{N}_i$, a predefined number q of the neighbors that provide it with the best intrusion detection and delete the others.

4. Go to the step 1.

This protocol contains a lot of freedom making it possible to be implemented in many particular options. In practice, it would be necessary to tune the protocol (e.g. on the step 3) and its attributes (e.g., cardinalities r, s, q) to particular application.

In fact, the described version of the protocol determines only its core remaining a number of options. In [10] it is demonstrated on the basis of a distributed networked intrusion detection case study when autonomous intrusion detection agents structured in p2p network implement p2p learning in a completely distributed way.

One can mention that all agents in the architecture presented in Fig. 1 are structured in p2p fashion and form several p2p overlay networks. In particular, this concerns not only sensor agent network, but also all other agents forming the smart space information field, implementing data, information and knowledge processing as well the agents accessing to web services. They solve many different tasks and have to form coalitions to effectively and efficiently solve every particular task. At the smart space installation time, all the application agents, the agents of white and yellow pages service, as well as the agents of p2p communication providers can be configured not in an appropriate way. The described distributed p2p learning algorithm (protocol) can be used to form the coalitions that are best tuned to particular tasks of the smart space thus supporting its adaptability, self–learning, etc.

4 Agent's Context– Driven Learning

It was emphasized in the paper introduction that one of the topmost tasks of the smart space R&D is providing, for smart space inhabitants, with context–dependent information, multiple services, reminders and personalized recommendations when and where necessary. Each such task should be solved via interaction of multiple agents. The quality and completeness of the extracted context is the topmost condition determining quality of provided services and their personalization the smart space is responsible for. Agents have to be learnt to extract context using modern achievements in context-dependent data mining. Moreover, smart space agents have to be capable to extract context-dependent features, detect causes determining smart space inhabitants' satisfaction by the services provided and recommendations done.

However data, information and knowledge available in smart space are distributed over many sources and stored in multiple data bases and therefore are difficult for on-line processing. Fig. 2 demonstrates the data and information sources available in smart space and their peculiarities implying the processing difficulties.

Operation of any smart space is accompanied by solution of many classification, recognition, prediction and decision making tasks. The most frequently task is recognition of human behavior patterns and actions to forecasting his/her subsequent behavior in order to recommend to him a service, which, according to the smart space ambient intelligence "opinion", could be useful in the current or in the forthcoming context [7]. For example, the simplest service is switching on and off different electrical devices when the user is moving through the rooms. The typical approach is monitoring of the user movements through the rooms and subsequent learning using logged data context. More complex task is recognition of the user typical behavior patterns and association them with the context. Many authors researching smart space emphasize an importance of the context for best fitting the user needs. In particular, the authors of the paper [7] use two attributes to specify context that are action start time and its duration. However it is found out that even such a primitive accounting of the context formalized in terms of Markov model is capable to significantly improve the predictive properties of the patterns detected. It is important to note that the authors of this research use agent and data mining technologies integration.

Let us consider an example where context plays the primary role, that is *Intelligent personal assistant* of MS Outlook e-mail. It can be a component of the smart space like *smart office* [8].

The history of a user work with the personal e-mailbox is reflected by the structure of the *Input* and *Sent* folders and the batches of e-mails stored in them, as well as by the mailbox holder contact list. The personal assistant objective is to recommend, for each input e-mail, the folder in which the latter has to be posted to.

The idea of context-dependent data processing is based on semantic interpretation of the input information based on rich ontology. Let us exemplify how the information contained in an e-mail can be significantly enriched by context extracted from high level knowledge contained in the domain ontology. Let us assume that the context to be extracted from an input e-mail of a particular user concerns with his/her business activity. An example of such e-mail is presented in Fig. 3.

One can see the diversity of information contained in this e-mail: e-mail *formal properties* (importance, sender and receiver, etc.), e-mail *subject*, *text* of e-mail

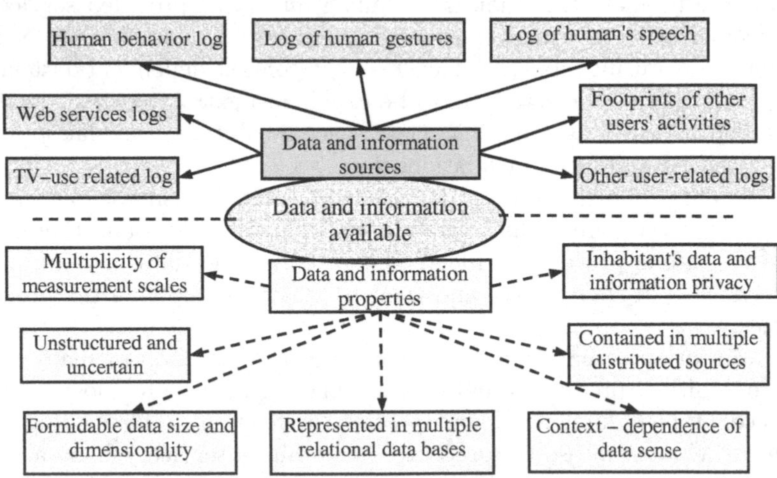

Fig. 2. Smart space data and information properties and sources

body with the associated re–writing history. Valuable information can also be extracted *indirectly* via joint analysis of the e-mail text and available information associated with persons' names mentioned in the e-mail body and subject.

The mentioned *indirect* information is exactly what is usually called as context, in our case – the context of the e-mail instance. Let us outline business activity associated domain ontology specifying the meta–level e-mail assistant knowledge. This ontology comprises two parts. The first one is intended to specify the basic high level concept *EMailItem* and its structure in terms of lower level ontology concepts. The second part of the ontology represents the second high level concept, *Person.* It is used to specify any person mentioned in the e-mail that is either the mail account owner or any other person mentioned in the contact list, e-mail subject or e-mail body. UML diagram of this ontology is depicted in Fig. 4.

It is important to note that the user's name *Person* (e-mail account owner) is the basic concept of the both parts of the ontology. *Person* may occupy a *Position* in a *Company* or several *Positions* in different companies, i.e. this relation is of $0\ldots1 - 0\ldots*$ cardinality. *Company* has its own postal address *PostAddress*, phone number *Phone*, web domain *WEBDomain* and web address *WebURL*, which are indirect properties of the e-mail account owner *Person.* *Person* is also characterized by *InstantMessengerUIN* list representing addresses (logins) of personal messengers and *nicknames* in social networks, e.g. addresses in Skype, ICQ, QIP, Yahoo-messenger, nickname in Facebook, Live Journal, etc. Every element of *Person*'s contact list *Contact* can, in turn, be connected to the concept *Person* since the people mentioned in the contact list *Contact*

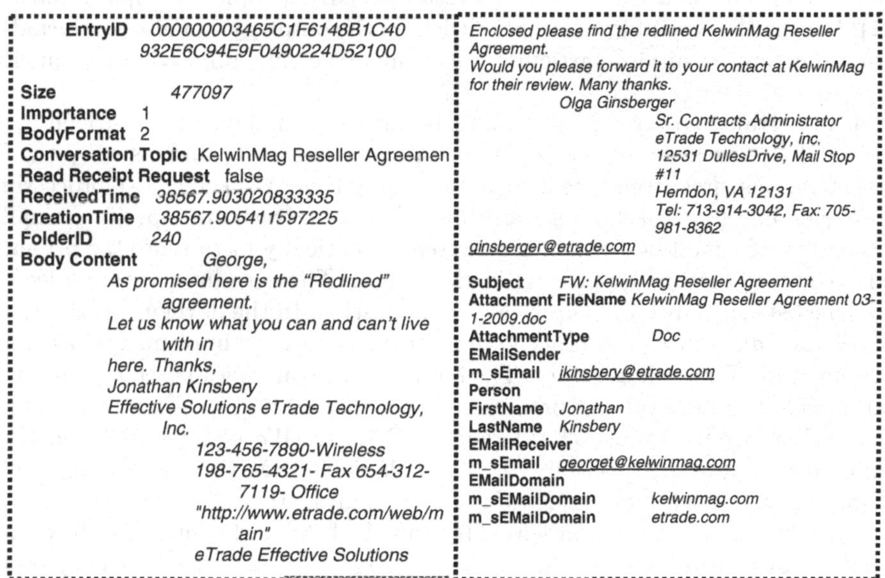

EntryID *000000003465C1F6148B1C40 932E6C94E9F0490224D52100*

Size *477097*
Importance 1
BodyFormat 2
Conversation Topic KelwinMag Reseller Agreemen
Read Receipt Request false
ReceivedTime *38567.903020833335*
CreationTime *38567.905411597225*
FolderID *240*
Body Content *George,*
As promised here is the "Redlined" agreement.
Let us know what you can and can't live with in
here. Thanks,
Jonathan Kinsbery
Effective Solutions eTrade Technology, Inc.
123-456-7890-Wireless
198-765-4321- Fax 654-312-7119- Office
"http://www.etrade.com/web/main"
eTrade Effective Solutions

Enclosed please find the redlined KelwinMag Reseller Agreement.
Would you please forward it to your contact at KelwinMag for their review. Many thanks.
Olga Ginsberger
Sr. Contracts Administrator
eTrade Technology, inc.
12531 DullesDrive, Mail Stop #11
Herndon, VA 12131
Tel: 713-914-3042, Fax: 705-981-8362
ginsberger@etrade.com

Subject *FW: KelwinMag Reseller Agreement*
Attachment FileName *KelwinMag Reseller Agreement 03-1-2009.doc*
AttachmentType *Doc*
EMailSender
m_sEmail *jkinsbery@etrade.com*
Person
FirstName *Jonathan*
LastName *Kinsbery*
EMailReceiver
m_sEmail *georget@kelwinmag.com*
EMailDomain
m_sEMailDomain *kelwinmag.com*
m_sEMailDomain *etrade.com*

Fig. 3. An example of e-mail instance

are described by the same schema as *Person* specifying *E-mail* account owner, i.e. by *Position,* etc. Thus, all the data associated with the contact list *Contact* can be imported in object database that is enriched with additional information about the set of instances of the ontology. This information can be stored in data bases that can contain, e.g., companies' names and descriptions of the companies and their products, some specific information about particular persons, etc.

Each *EMailItem* concept instance is assigned a set of attributes. Among them, some are *formal* ones specifying relations between *EmailItem* and *Email.* The e-mail properties associated with the e-mail body *BodyContent* specify most informative *informal* properties *of EMailItem* that are the basic source of contextual information. Specific attribute of e-mail instance is *the name of folder in which e-mail is posted.* For every new e-mail, this attribute is unknown: it is the subject of the decision to be produced by the intelligent e-mail assistant in question on the basis of the available information and its fusion. E-mail is also described by its subject *EMailSubject*; it may be attached with one or more files *Attachment* having attribute *AttachmentType.*

Other relations (Fig. 4) can also contain contextual information. It can be formalized in terms of *secondary concepts* to be introduced by an expert to specify both the body of *EmailItem* instance and data from contact list *Contact.*

Secondary concepts determined by an expert are to be included in the ontology as well. E.g., the name of a company may become a secondary concept if its phone number is found in the e-mail body. In such a case, the company name is considered as a concept *connected to the e-mail body.* Some of them are auxiliary (e.g. *RelatedItems* of *BodyContent*, *EMailSubject, Attachment*, etc.), whereas other are to be involved in data mining process (e.g. *key words*, *e-mail addresses*, *phone numbers*, instances of *InstantMessengerUIN*, people's/company *names*, etc.). In fact, the secondary concepts enriching the ontology can also significantly enrich the context of any particular e-mail instance thus improving the quality of decision making.

An important part of information to be involved in the context–driven data mining of agents is contained in the e-mail message body and its subject. To extract the features from a text in natural language, a specific text processing techniques has to be used. They are intended to extract and specify formally the context of e-mail body and subject to automatically instantiate the primary and secondary features appearing in the e-mail. These techniques are selected and tuned specifically to the application in question. In the developed *Intelligent e-mail assistant* software prototype these techniques were implemented as components of MS Outlook assistant. The details of the ontology-based e-mail body text as well as e-mail subject processing can be found in [8,9]. These techniques are based on two IBM tools: *IBM Language Resource Ware* (LRW) [13] and *IBM Ontological Network Miner* (ONM) [14] available at IBM's alphaworks site [20].

The result of such transformation of an e-mail instance is an object (in the sense of *object data base*) represented in star-like form and containing the entire context that can be extracted from the ontology. The sample of such representations of e-mails of the structured mailbox folders is further used as learning

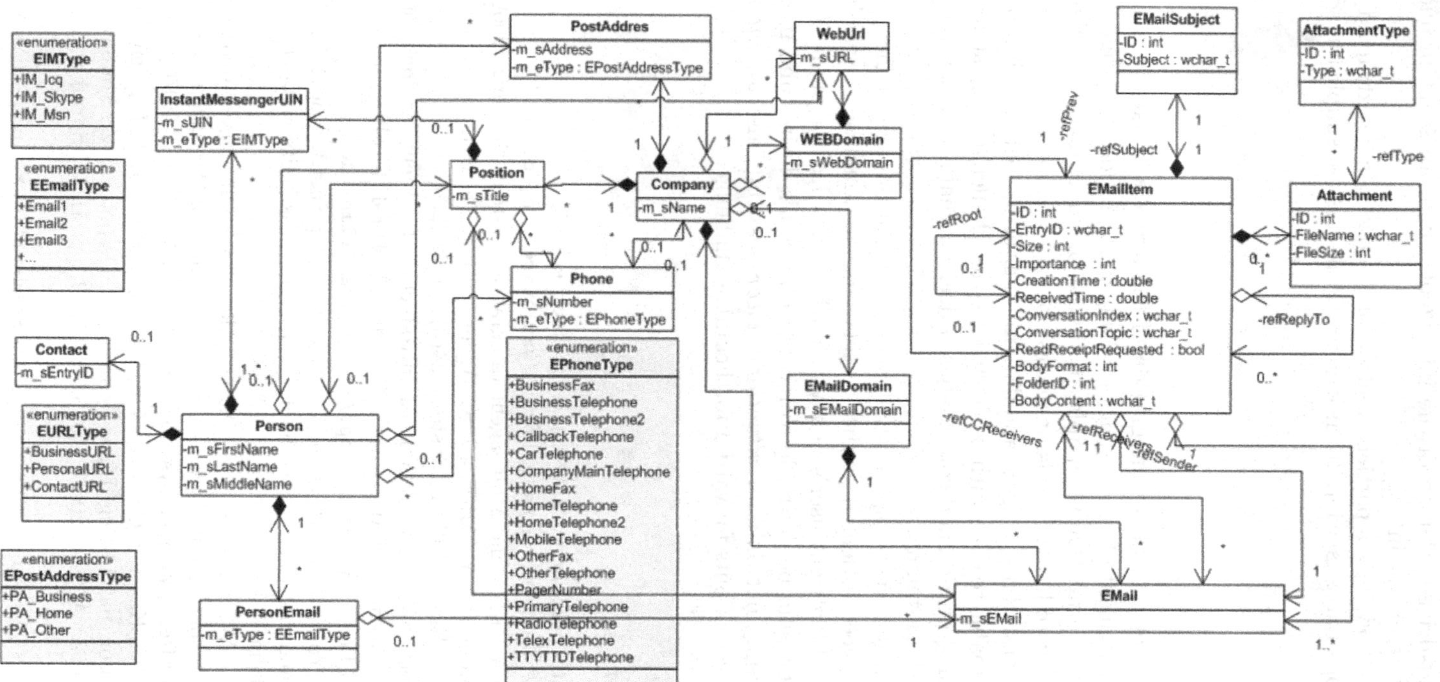

Fig. 4. User's business activity associated domain ontology of an e-mail assistant

data for context-driven training of the intelligent e-mail assistant in question. The details are described in [8].

The example shows how a particular situation context can be extracted, specified and used for smart space intelligence enrichment. It is not simple but solvable task.

5 Conclusion

The paper analyses the role of integration of the agent and data mining technologies in management of smart space challenges. It proposes p2p agent-based smart space architecture that is capable to provide for smart space with flexibility needed to adapt it to human needs: it makes applicable, to smart space, modern advances in integrated use of agent and data mining technologies. This integrated technology is capable to cope with basic smart space challenges, e.g. smart space software self-configuration and context-based decision making achieved through context-driven data mining.

Nevertheless, new smart space challenges are forthcoming. In fact, to provide for human needs-oriented predictive servicing, it would be very helpful to recognize not only current user's behavior patterns needed for predictive decision making. It would be desirable to recognize human's desires and intentions too, thus providing not only for *situational awareness*, but also for *intentional awareness*. This task puts forward a forthcoming challenge.

References

1. Angel.Tb site, `http://www.ist-angel-project.eu`
2. AWARE. FP-6 Project web site, `http://www.aware-project.net/`
3. Cao, L.: Data Mining and Multi-agent Integration (edited). Springer (2009)
4. Cao, L., Gorodetsky, V., Mitkas, P.A.: Agent Mining: The Synergy of Agents and Data Mining. IEEE Intelligent Systems 24(3), 64–72 (2009)
5. Cao, L., Weiss, G., Yu, P.S.: A Brief Introduction to Agent Mining. Journal of Autonomous Agents and Multi-Agent Systems 25, 419–424 (2012)
6. CobiS. FP-6 Project web site, `http://www.cobis-online.de`
7. Cook, D.J., Youngblood, G., Jain, G.: Algorithms for Smart Spaces. In: The Engineering Handbook of Smart Technology for Aging, Disability and CityplaceIndependence. John Wiley & Sons, Inc. (2008)
8. Gorodetskiy, V., Samoilov, V., Serebryakov, S.: Ontology–based Context–dependent Personalization Technology. In: International Workshop "Web Personalization and Recommender Systems" at IEEE/ACM WI/IAT 2010 (2010)
9. Gorodetsky, V., Karsaev, O., Samoylov, V., Serebryakov, S.: P2P Agent Platform: Implementation and Testing. In: Joseph, S.R.H., Despotovic, Z., Moro, G., Bergamaschi, S. (eds.) AP2PC 2007. LNCS, vol. 5319, pp. 41–54. Springer, Heidelberg (2010)
10. Gorodetsky, V., Karsaev, O., Samoylov, V., Serebryakov, S.: Interaction of Agents and Data Mining in Ubiquitous Environment. In: IAT-ADMI 2008, pp. 9–12 (2008)

11. Gorodetsky, V., Karsaev, O., Samoylov, V., Serebryakov, S.: P2P Agent Platform: Implementation and Testing. In: Joseph, S.R.H., Despotovic, Z., Moro, G., Bergamaschi, S. (eds.) AP2PC 2007. LNCS, vol. 5319, pp. 41–54. Springer, Heidelberg (2010)
12. Hydra. FP-6 Project web site, `http://www.hydra.eu.com/`
13. IBM Language Resource Ware, `http://www.lphaworks.ibm.com/tech/lrw`
14. IBM LanguageWare Miner for Multidimensional Socio-Semantic Networks, `http://www.alphaworks.ibm.com/tech/galaxy`
15. MORE. FP-6 Project web site, `http://www.ist-more.org`
16. Runes. FP-6 Project web site, `http://www.ist-runes.org`
17. Sense. FP-6 Project web site, `http://www.sense-ist.org`
18. SMEPP. FP-6 Project web site, `http://www.smepp.org`
19. SOCRADES. FP-6 Project web site, `http://www.socrades.eu`
20. Text Analytics Tools and Runtime for IBM Language Ware, `http://www.alphaworks.ibm.com/tech/lrw`
21. Jih, W., Hsu, J.Y., Lee, T., Chen, L.: A Multi-agent Context-aware Service Platform in A Smart Space. Journal of Computers 77(3), 45–60 (2007)

Agent-Mining of Grid Log-Files: A Case Study

Arjan J.R. Stoter[1], Simon Dalmolen[1,2], and Wico Mulder[1]

[1] Logica, Amstelveen, The Netherlands
{arjan.stoter,simon.dalmolen,wico.mulder}@logica.com
[2] School of Management & Governance,
University of Twente Enschede, The Netherlands

Abstract. Grid monitoring requires analysis of large amounts of log files across multiple domains. An approach is described for automated extraction of job-flow information from large computer grids, using software agents and genetic computation. A prototype was created as a first step towards communities of agents that will collaborate to learn log-file structures and exchange knowledge across organizational domains.

Keywords: Grid monitoring, text mining, agent oriented programming, genetic computation, engineering.

1 Introduction

Grid infrastructures are distributed and dynamic computing environments that are owned and used by a large number of individuals and organizations. The EGEE [1] grid for instance builds on the collective efforts of over 140 organizations in 50 countries, where each organisation owns and manages part of the grid. In such environments, computational power and resources are shared to process jobs. A job is a computation task launched by a client to be handled by the grid. This job is pointed to a resource by a scheduler and then executed on multiple resources in different parts of the grid, supported by different cluster organizations.

Operational management of grids is challenging. There are many components and interactions, resources may join and leave any time, resources are heterogeneous and distributed across organizations, and the components undergo continuous improvements and changing of standards [2]. The collaborative and often dynamic settings of grids requires an intelligent information management system that is not only flexible and extensible but also able to relate information from different organizational domains.

A rich source of information for managing grids are system log-files. Log files can be used for a number of things, such as performance analysis, security management, and user profiling [3]. In grids, log files are used to analyse network paths and job flows.

Today, system managers in grids use a variety of tools to monitor the status of the grid, such as Monalisa, Nagios, BDII, and RGMA. However, these tools

L. Cao et al.: ADMI 2012, LNAI 7607, pp. 166–177, 2013.

are typically restricted to the boundaries of a single organizational domain. An organisational domain in this case refers to a cluster organisation that owns and maintains a hardware cluster of the grid.

In practice, grid log-files are often analysed manually by local domain administrators. The main reason for this is that the log-files not only contain information about jobs but also other entries related to for instance system performance and security. Job-flow analysis therefore requires specific knowledge about the relevant entries related to job processing and the way in which they are structured. This makes job-flow analysis a complex task for administrators, especially when administrators from different domains have to work together to retrace errors in job processing [2].

Figure 1 illustrates collaboration between grid administrators. Each domain contains log files (manifests) with sometimes different structures. Retrieving the executed path of a job and finding the reason of failure is done manually by domain administrators. Using his or her domain knowledge, administrators scan the manifests for regions of interest (ROIs). An ROI is, for example, an IP address or a unique job identifier. Some administrators make use of regular expression to represent ROIs that match log entries related to jobs. He or she then manually creates an information extraction pattern to extract rich information from the manifests. This information extraction pattern is specific to his or her domain due to the differences in log-file structures across domains. Defining ROIs and domain specific patterns for information extraction is challenging and makes error tracing in grids a complex and time-consuming task, which often requires communication between domain administrators to exchange their domain knowledge to achieve a cross domain overview.

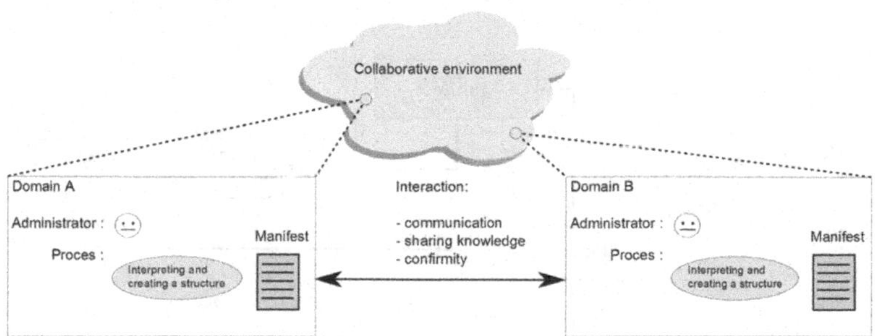

Fig. 1. Schematic representation of collaboration between grid administrators in different domains

2 Approach

Error tracing in grids could be optimized profoundly if administrators would be able to automatically exchange domain knowledge and then automatically incorporate this acquired knowledge into a new and improved pattern for information-extraction from the manifests within their domain.

In the current study, a prototype was created that integrated methodologies from multi-agent technology [4], data mining, and knowledge discovery, known as agent mining [5]. Agent mining has shown a high potential to enhance collaboration performance and tackling of errors and exceptions in distributed computing environments, such as grids and clouds [6].

The current prototype incorporated techniques from agent-based data mining, where agents collaborated inside a multi-agent system (MAS) to optimize information retrieval and gathering of information across distributed nodes in the grid by means of distributed learning [7]. Each agent inside the MAS located and refined ROIs inside log files using domain knowledge and created a domain specific pattern for information extraction, called a corpus. The term corpus stems from text mining, a research area that focuses on the identification and extraction of relevant features inside manifests. In text mining, unstructured or semi-structured data is first transformed into a structured intermediate format. This structured format -or corpus- can then be queried to extract data from manifests [8]. By exchanging their knowledge, the agents learned from one another to achieve higher levels of data abstraction. The goal of the current study was to investigate whether such an approach has the potential to support domain administrators in grids during error tracing.

Figure 2 shows the scope of the prototype. Manifests containing log data from actual grids were used as data. During preprocessing, initial domain knowledge was acquired about the manifests based on a priori knowledge from system administrators. This knowledge consisted of domain knowledge and structure knowledge. Domain knowledge contained a list of attributes that the agent assumed to be of relevance in the log file, the ROIs. These are the data patterns the agent would search for in the log file. These attributes were called descriptive

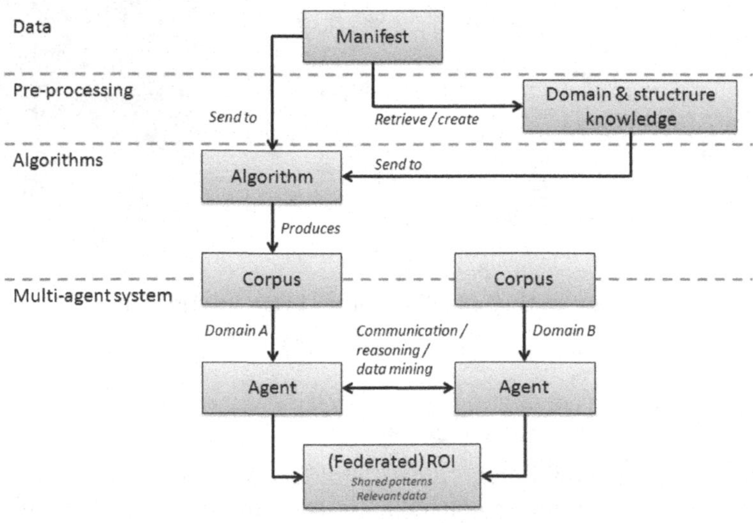

Fig. 2. Scope of the prototype

attributes (DA). DAs were represented by regular expressions, as shown in table 1. Structure knowledge contained a collection of concepts that serve as building blocks to construct the regular expressions.

Table 1. Examples of DAs and ROIs in domain knowledge

DA/ROI	Log-file line example
TIME\pPunct*\s*({\w+ \s*:}*)	TIME: Sun Dec 7 04:02:09 2008
PID\pPunct*\s*(\w1,)	PID: 125690

Domain- and structure knowledge were used by an algorithm to locate the ROIs within the manifest, construct a regular expression to represent each ROI, and create a corpus. Finally, the constructed ROIs were federated between the software agents.

Genetic computation was used to construct the corpus. Genetic algorithms use Darwin's principle of natural selection, along with analogs of recombination (crossover), mutation, gene duplication, gene deletion, and mechanisms of developmental biology [9]. A genetic algorithm follows an evolutionary path towards a solution. These solutions are represented by chromosomes made up of individual elements called genes. Genes encode specific kinds of information. Populations of chromosomes (i.e. possible solutions) evolve into new generations of chromosomes by means of recombination (crossover) and mutation. A genetic algorithm typically has the following logic [10]:

1. Create a population of random chromosomes.
2. Test each chromosome for how well it fits the solution.
3. Assign each chromosome a fitness score.
4. Select the chromosomes with the highest fitness scores and allow them to survive to a next generation.
5. Create a new chromosome by using genes from two parent chromosomes (crossover).
6. Mutate some genes in the new chromosome.

Steps 2 to 6 are repeated until a certain condition is met, or when a maximum number of generations is reached.

After construction of the corpus, the agents federated domain knowledge to other agents inside the MAS, who in turn incorporated these new ROI definitions inside their domain knowledge and used them to create a new corpus to read manifests. The JADE framework was used to create a MAS. JADE is an Open-Source, Java-based framework, and one of the most widespread agent-oriented middleware in use today [11].

3 Prototype

Figure 3 shows a schematic representation of the scope of a single agent. Each element of the figure is explained in the next paragraphs.

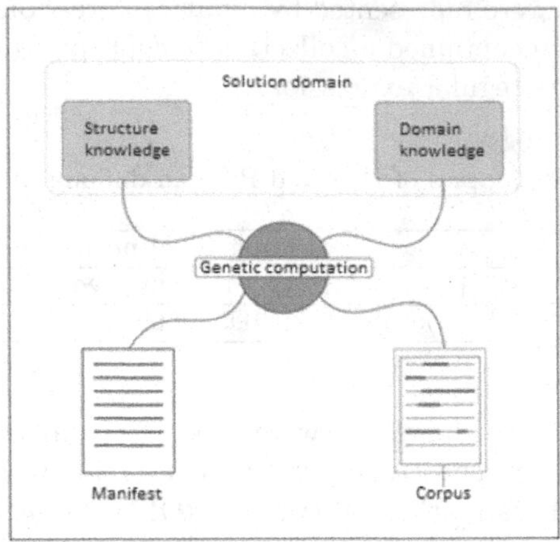

Fig. 3. Representation of the elements within the scope of a single agents

The used data were manifests that contained original log data from the EGEE grid, provided by the National Institute for Nuclear and High energy physics (NIKHEF).

The solution domain in figure 3 represents the agent's pre-knowledge of the manifest: its assumptions about the manifest's structure and interesting parts. The solution domain consists of domain knowledge and structure knowledge. Domain knowledge and structure knowledge together provide the building blocks for an agent to construct a corpus. The corpus was a collection of ROIs ordered in a sequence that has the highest fit on a reoccurring pattern in the manifest.

The solution of genetic computation was a chromosome that represented a corpus that fitted a data pattern inside the manifest. The chromosomes consisted of elements from structure knowledge and domain knowledge, which formed the genes within the chromosome. Each chromosome represented a potential corpus. The size of chromosome was a fixed size, which was set before run-time. The corpus was represented by a directed graph as shown in figure 4. Directed graph G is

$$G = (N, E)$$

where N is the set of nodes and E is the set of edges. The edge set E is a subset of the cross product $(N * N)$. Each element (u,v) in E is an edge joining node u to node v. A node v is neighbour of node u if edge (u,v) is in E.

An $(N * N)$ adjacency matrix (table 2) represented the adjacency of vertices within the corpus. So a chromosome consisted of multiple adjacency-matrix index numbers to represent the graph, and the index numbers referred to a directed edge where the node was an ROI. Index number 1 meant node 2 is connected to node 1 where node 2 is the parent node of node 1. A root node was defined as

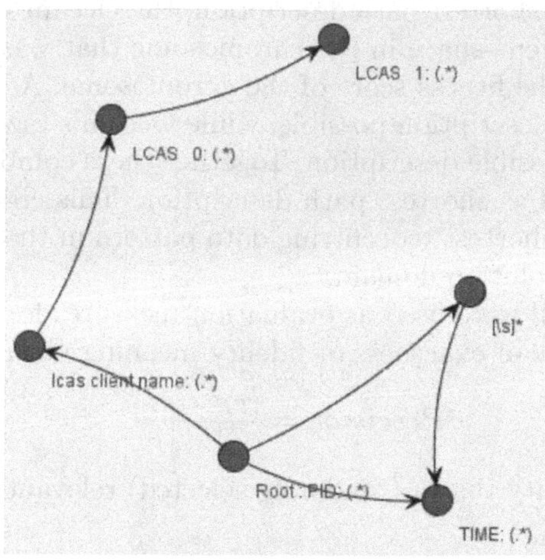

Fig. 4. Example of a directed graph, representing a corpus

a starting point for the corpus (in this case PID). In sum, first a root node was defined as a starting point for the graph. Next, the adjacency index numbers were translated into a graph of regular expressions.

Table 2. Example of an adjacency matrix. Parent node 2 has child node 3.

u / v	Node 1	Node 2	Node 3
Node 1	0	1	2
Node 2	3	4	5
Node 3	6	7	8

The genetic algorithm fitted each graph (chromosome) onto the manifest, where the root node was taken as the starting point of the graph. After a fit of the root node it tried to fit the first child node on each following line of the manifest. After a match, the next child node was fitted, and so on. This recursive process was repeated until the function reached the end of the manifest or until the root node fitted again. The latter was taken as a re-occurrence of the pattern.

A function *ScoreGraph* was used to return the longest path, or fit of nodes, between two root nodes. The *ScoreGraph* function returned the best path including recursion of the directed graph (chromosome), and was defined as

$$ScoreGraph = argmax(followedpaths)$$

When the whole manifest was covered and scored -by fitting the graph- the total score was summed. A fitness score was calculated, where n was the fit of a root node, with $k = 1$ summing the scores of *ScoreGraph*.

$$FitnessScore = \sum_{k=1}^{n} ScoreGraph^{k}$$

For optimizing to the shortest path description, the Occam's razor [12] principle was used. For every gene-space in the chromosome that was not filled, a bonus of 0.1 was added to the fitness score of the chromosome. As such, argmax tried to create the longest description possible, while Occam's razor principle tried to create the shortest possible description. Together these combined forces resulted in a complete, as well as shortest path description. This created a chromosome that represented the shortest reoccurring data pattern in the manifest, given the knowledge from the solution domain.

Precision and recall were used as evaluation measures for validating an ROI. Precision is a measure of exactness or fidelity meaning nothing but the truth:

$$Precision = \frac{|TP|+|FP|}{|TP|}$$

Recall is the probability that a (randomly selected) relevant item is retrieved in a search.

$$Recall = \frac{|TP|}{|TP|+|FN|}$$

The true positives (TP) represented the number of detected correct matches, false positives (FP) were the number of falsely detected matches, false negatives (FN) represented the number of unidentified data that were in fact real matches, and true negatives (TN) were the number of unidentified data that were not matches.

First, the genetic algorithm was tested using the manifest as input data, which contained 524 lines of entries. The amount of TPs within the manifest was known before hand, and consisted of 151 TPs.

The fitness function used a random mutation operator. A random mutation operator means a new random mutation rate was created for each evolution and applied to the current generation of chromosomes. Chromosome size and population size were manipulated. Six configurations were tested, and al configurations were tested 5 times (runs) over 300 evolutions. In this case the mutation operator was randomized every evolution:

1. Chromosome size = 11 (equal to the ultimate adjacencies), population size = 50
2. Chromosome size = 11 (equal to the ultimate adjacencies), population size = 100
3. Chromosome size = 15, population size = 50
4. Chromosome size = 15, population size = 100
5. Chromosome size = 20, population size = 50
6. Chromosome size = 20, population size = 100

3.1 Multi-agent System

The prototype was tested using several agents that had limited knowledge about a log file. The goal of the prototype was to show that agents could learn from one another by exchanging domain knowledge and structure knowledge. In the current setting, four agents were given the same log file to analyse while holding a different set of DAs.

Each agent was given the location of the manifest and contained a JADE behaviour to analyse the manifest. The genetic algorithm was implemented in the JADE agents. The prototype uses a minimal vocabulary (JADE ontology) for communication between agents, which consisted of *corpusscore*, *score*, *da*, *name*, *expression*, *ihavescore*, and *ihaveda*.

CorpusScore was the internal fitness score of each agent. *Score* was the score received from other agents. *DA* was an ROI, *name* was the name of the DA, and *expression* was the regular expression to represent the DA. *IhaveScore* was an attribute that indicated that an agent had a corpusScore, and finally *IhaveDA* was an attribute to indicate than an agent had one or more DAs.

Figure 5 shows the interaction between two agents. The prototype was tested using four agents located on two different machines. Each agent had different DAs but all had the same root node. Each agent built a corpus every 5 minutes and exchanged DAs according to the sequence diagram in figure 5. After an agent was created it first located and opened the manifest. Next, it created a corpus using genetic computation and its solution domain. When an agent successfully created a corpus with a fitness score on the manifest, the agent would broadcasts its fitness score to the other agents. When the incoming score (received from other agents) exceeded its own, the agent would broadcast a request for DAs. The agent with the highest score would respond with a *IhaveDA* that contained its DAs. Next, the requesting agent would add the DAs to its domain knowledge.

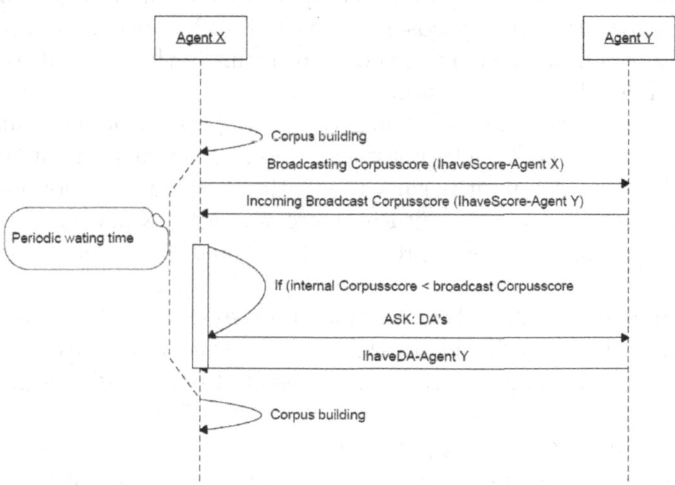

Fig. 5. Sequence diagram agent mining

4 Results

Table 3 shows the results of genetic computation with a random mutation. The max hit score was 151, and the bonus score for the shortest description was added to find the optimal corpus. In the case of chromosome size 20 the optimal score was 151.9 because 9 adjacencies should then be empty within the chromosome.

Results show that when the chromosome size was equal to the size of the optimal adjacencies (in this case 11) the optimal corpus was not found. A chromosome length of 15 required a lower number of generations to reach the optimal corpus than a chromosome size of 11 and 20. The precision score was always 1 in this test. This was because of the fact that ROIs were implemented as regular expressions. Regular expressions have a hard fit, which decreases the chance of FPs.

Two out of six configurations never reached the optimal corpus: chromosome size 11 with a population of 50 and 100. The parameter configuration of chromosome size 11 and population size 50 reached 144.1 two times. So the corpus graph score had 144 fit points, and one allele in the chromosome remained empty. This resulted in the additional bonus score of 0.1. The optimal score was 151 and all the alleles in the chromosome should be filled. Results show that the maximum fitness score was reached at an average of 90 evolutions, with an average fitness score of 140. With a population size of 100 and chromosome size of 11, the average highest fitness score was 141.3, which was reached after an average of 25 generations. It suggested that a higher population size resulted in a higher fitness score compared to a population of 50. In addition, the highest score was also reached at an earlier stage of evolution. With a population of 50, an average of 140 evolutions was needed to reach the highest score, and with a population of 100 the result was an average of 25 evolutions. A population of 100 reached its maximum score 115 evolutions sooner, compared to a population of 50. These results suggest that using a chromosome size equal to its optimal adjacencies (in this case 11, known from the manifest) did not result in the optimal corpus, and tended to reach local optimal solutions.

Using the parameters of chromosome size 15 and population 50 resulted in an average evolution number of 116 before reaching its average maximum fitness score of 150. During 4 out of 5 runs the optimal corpus was found. With a population size 100, the average fitness score was approximately 149 and the average evolution number was approximately 40, and 3 out of 5 runs reached the optimal corpus.

A chromosome size of 20 and population of 50 resulted in the optimal corpus during all runs of the experiment. In this case the average evolutions (resulting in a maximum score) was 201.6. Both cases resulted in a recall and precision of 1. When the population was set to 100 the average of evolutions that reached its maximum fitness score was approximately 99.

Results suggested that doubling the chromosome size reduced the chance of local optima solutions. Also, a higher starting population seemed to reduce the number of evolutions required to reach the optimal corpus. In this experiment chromosome size 20 and population 100 showed the best result for acquiring the optimal corpus (without respect to computation time).

Table 3. Genetic computation result with a random mutation

Chromosome size	Population size	Run	Start fitness evolution at 1	Max. fitness score	Max fitness score at evolution	Ultimate corpus found	Precision	Recall
11	50	1	79.0	137.2	37	NO	1	0.907
11	50	2	99.1	144.1	101	NO	1	0.954
11	50	3	72.0	130.3	23	NO	1	0.861
11	50	4	89.0	137.1	261	NO	1	0.907
11	50	5	99.0	144.1	25	NO	1	0.954
15	50	1	96.0	151.4	56	YES	1	1
15	50	2	85.1	151.4	76	YES	1	1
15	50	3	106.0	151.4	94	YES	1	1
15	50	4	96.0	151.4	279	YES	1	1
15	50	5	106.0	144.5	73	NO	1	0.954
11	100	1	99.0	144.1	18	NO	1	0.954
11	100	2	89.0	144.1	26	NO	1	0.954
11	100	3	89.1	144.1	21	NO	1	0.954
11	100	4	106.0	144.1	30	NO	1	0.954
11	100	5	99.0	130.3	27	NO	1	0.861
15	100	1	106.0	151.4	32	YES	1	1
15	100	2	113.0	151.4	28	YES	1	1
15	100	3	99.0	144.5	65	NO	1	0.954
15	100	4	99.0	144.5	38	NO	1	0.954
15	100	5	113.0	151.4	35	YES	1	1
20	50	1	106.0	151.9	274	YES	1	1
20	50	2	106.0	151.9	225	YES	1	1
20	50	3	103.1	151.9	197	YES	1	1
20	50	4	113.0	151.9	200	YES	1	1
20	50	5	96.0	151.9	112	YES	1	1
20	100	1	113.0	151.9	121	YES	1	1
20	100	2	116.1	151.9	95	YES	1	1
20	100	3	113.0	151.9	132	YES	1	1
20	100	4	120.1	151.9	63	YES	1	1
20	100	5	116.1	151.9	82	YES	1	1

The prototype illustrated that agents were able to learn from each other. Over time each of the four agents was able to build identical and optimal corpora, because they shared all the DAs known in the agent network.

5 Discussion

A MAS prototype was created to exchange domain knowledge between agents. This exchanged knowledge was then used by each agent to create a corpus for information retrieval from log-file manifests. The aim of the prototype was to investigate whether such an approach has the potential to support domain administrators in grids during error tracing.

The created prototype was able to successfully exchange knowledge between agents. Each agent was then able to use this knowledge to create an optimal corpus for information retrieval. The prototype allowed extraction of data from log-files even when the structure of the log-files changes over time, and new or changed domain knowledge can be introduced and shared easily among agents. The prototype therefore illustrated collaborative learning and the automatic integration of knowledge for reading manifests, and showed the potential to support information retrieval in a cross-domain volatile environment.

The prototype represented a first step towards communities of agents that will collaborate to learn log-file structures and exchange knowledge across organizational domains. Some limitations of the prototype will have to addressed in follow-up research. For instance, the current prototype was successful when analysing log files that were similar in structure and contained similar ROIs. Future work will have to address comparison of log-files in different cluster organisations, which contain different entry orders as well as ROIs that are different in nature. The latter would require semantic translation and comparison between domain knowledge of cluster organisations. For instance PID in one cluster domain may be represented as ID in another. The matching of ROIs on a semantic level was outside the scope of the current prototype.

Finally, the current study did not address scalability, deployment, and overhead of the prototype in a production cluster. While the JADE framework is known for its scalability and deployment in distributed systems [11] the genetic computation that was used could potentially increase overhead. Impact on system performance and optimization should be addressed in future work.

Acknowledgements. This study was performed within Collaborative Network Solutions (CNS), an expertise group at Logica, in collaboration with NIKHEF. The study was funded by the VL-e project (www.vl-e.org).

References

1. EGEE: EGEE Homepage, `http://public.eu-egee.org/`
2. Mulder, W., Jacobs, C.: Grid management support by means of collaborative learning agents. In: Proceedings of the 6th International Conference Industry Session on Grids Meets Autonomic Computing, pp. 43–50. ACM (2009)

3. Oliner, A., Ganapathi, A., Xu, W.: Advances and challenges in log analysis. Communications of the ACM 55, 55–61 (2012)
4. Russell, S., Norvig, P.: Artificial Intelligence: A modern approach, 3rd edn. Prentice-Hall, New Jersey (2009)
5. Cao, L., Gorodetsky, V., Mitkas, P.A.: Agent Mining: The Synergy of Agents and Data Mining. IEEE Intelligent Systems 24(3), 64–72 (2009)
6. Cao, L.: Data Mining and Multi-agent Integration (edited). Springer (2009)
7. Cao, L., Weiss, G., Yu, P.S.: A Brief Introduction to Agent Mining. Journal of Autonomous Agents and Multi-Agent Systems 25, 419–424 (2012)
8. Feldman, R., Sanger, J.: The text mining handbook: advanced approaches in analyzing unstructured data. Cambridge University Press (2007)
9. Koza, J.R., Keane, M.A., Streeter, M.J., Adams, T.P., Jones, L.W.: Invention and creativity in automated design by means of genetic programming. Artificial Intelligence for Engineering Design, Analysis and Manufacturing 18, 245–269 (2004)
10. Conrad, E.: Detecting Spam with Genetic Regular Expressions. SANS Institute Reading Room (2007), http://www.giac.org/certified_professionals/practicals/GCIA/0.793
11. Bellifemine, F.L., Caire, G., Greenwood, D.: Developing multi-agent systems with JADE. Wiley (2007)
12. Blumer, A., Ehrenfeucht, A., Haussler, D., Warmuth, M.K.: Occam's razor. Information Processing Letters 24, 377–380 (1987)

A Proficient and Dynamic Bidding Agent for Online Auctions

Preetinder Kaur, Madhu Goyal, and Jie Lu

DeSI lab, Centre for Quantum Computation and Intelligent Systems
School of Software, University of Technology, Australia
preetinder.kaur@student.uts.edu.au,
{madhu,jielu}@it.uts.edu.au

Abstract. E-consumers face biggest challenge of opting for the best bidding strategies for competing in an environment of multiple and simultaneous online auctions for same or similar items. It becomes very complicated for the bidders to make decisions of selecting which auction to participate in, place single or multiple bids, early or late bidding and how much to bid. In this paper, we present the design of an autonomous dynamic bidding agent (ADBA) that makes these decisions on behalf of the buyers according to their bidding behaviors. The agent develops a comprehensive method for initial price prediction and an integrated model for bid forecasting. The initial price prediction method selects an auction to participate in and then predicts its closing price (initial price). Then the bid forecasting model forecasts the bid amount by designing different bidding strategies followed by the late bidders. The experimental results demonstrated improved initial price prediction outcomes by proposing a clustering based approach. Also, the results show the proficiency of the bidding strategies amongst the late bidders with desire for bargain.

Keywords: Online auctions, Software agents, Bid forecasting, Bidding strategies, Data mining, Clustering.

1 Introduction

The online auctions have received an extreme surge of popularity in the past few years. In the online auctions' business model, traders purchase or sell products governed by specific trading rules over the internet supporting different auction formats. There are four standard online auction formats: English auction; Dutch auction; First-price sealed-bid; and Second-price sealed-bid [1] [2]. English auctions are the most common auction type employed by the online auctioneers in eBay, Amazon etc. Bidders participating in this marketplace often face a challenge to opt for the most favorable bidding strategies to win the auction. Moreover, there are always a number of multiple auctions selling the desired item at the same time. This foster the complicated situation of the bidders in making decisions of selecting which auction to participate in, placing single or

L. Cao et al.: ADMI 2012, LNAI 7607, pp. 178–190, 2013.

multiple bids, bidding early or late and how much to bid [3] [4]. These hard and time consuming processes of analyzing, selecting and making bids and monitoring are needed to be automated to assist the buyers while bidding. Software agents can promisingly act upon these tasks on behalf of the traders. These are the software tools that can execute autonomously, communicates with other agents or the user and monitors the state of its execution environment effectively. These negotiating agents outperform their human counterparts because of their systematic approach to execute complex decision making situations [5] [6]. The software agents make decisions on behalf of the consumer and endeavor to guarantee the delivery of the item according to the bidder's behavior.

In eBay auctions, a bidder with the highest value wins and he pays the second-highest maximum price. However, the bidders do not bid their maximum values either because they have trouble understanding that they should bid their maximum value or they have trouble simply figuring out what their maximum value is. Closing price prediction of the auction can help the bidders in setting their maximum valuation of the auctioned item. Furthermore, the bidders adjust their bids towards the maximum valuation of the item repeatedly in response to the remaining time left of the auction and the bids placed by the other participants, which leads to the different bidding behavior. According to different bidding behaviors, we can categorize agents as evaluators, participators, opportunists, skeptics, snipers, unmaskers or shill bidders [7] [8] [9]. Moreover, late bidding in online auctions has aroused a good deal of attention [10] [11] [12]. Late bidders appear near to the closing hours of an auction. This is a very common behavior by the bidders in an eBay style auction with hard closing rules. Late bidding may be a best response to a variety of incremental bidding strategies because they can better realize the state of the auction by observing its historical data. So there is a need to design a mechanism which decides the bid amount at particular moment of time according to the bidding behavior of the late bidders.

Predicting the closing price of an online auction is a challenging task because it depends on auction's attributes which are dynamic in nature [13]. This bid amount can be forecasted effectively by analyzing the data produced as an auction progresses (historical data). Analysis of plethora of data produced in the online auction environment can be done by using data mining techniques to predict the closing price of an online auction [14] [15]. Data from a series of the same or similar auctions closed in the past has been used to forecast the winning bid by exploiting regression, classification and regression tree, multi-class classification and multiple binary classification tasks [16] [17]. Also the history of an ongoing auction contains significant information and is exploited for the short term forecasting of the next bid by using support vector machines, functional k-nearest neighbor, clustering, regression and classification techniques [13] [15] [18]. However, these forecasted bids have hardly been used to improve the behavior of the software agents in online auctions [19] [20].

This article develops an automated dynamic bidding agent (ADBA) that will use the data mining techniques to improve its behavior for bid forecasting according to different bidding strategies followed by the late bidders. ADBA selects an auction to participate in and predicts its closing price by adopting a clustering, and bid mapping and selection approach. Then the bidding agent models the bid amount for a given behavior based on the predicted closing price of the selected auction and the negotiation decision functions by Faratin et al [21]. These negotiation decisions are made based on the strategies generated by the bidding agent following different behaviors of the late bidders.

The rest of the paper is organized as follows. In section 2 we present the design of the Automated Dynamic Bidding Agent. Section 3 depicts experimental results evaluating the performance of the closing price prediction method and the success rate of the bidding strategies designed for different bidding behaviors of the late bidders. Section 4 concludes the paper.

2 Automated Dynamic Bidding Agent-ADBA

The bidding agent consists of two primary components: an initial price estimator, and a bid forecaster. The initial price estimator is responsible for the selection of an ongoing auction for participation and for predicting the closing price of the selected auction. The predicted closing price of the selected auction acts as the initial price for the bid forecaster. The bid forecaster utilizes this initial price for forecasting the bid amount for the selected ongoing auction based on different bidding strategies followed by the late bidders. The automated dynamic bidding agent (ADBA) is represented in Fig. 1.

2.1 Initial Price Estimator

The initial price estimator selects an auction to participate and predicts the closing price of the selected auction based on a clustering, and a bid mapping and selection approach. Formally our approach consists of four steps. First, historical data is extracted as per the requirements to form the agents' knowledge base for online auctions. Secondly, $k - estimator$ agent determines the best number of partitions for the overall auction data. Thirdly, the set of similar auctions are clustered together in k groups and finally, based on the transformed data after clustering and the characteristics of the current auctions, bid mapping and selection component nominates the cluster for each of the ongoing auctions to select the auction for participation, and the closing price of the selected auction will be predicted.

Data Preprocessing. Let A be the set of the attributes collected for each auction then $A = \{a_1, a_2, ..., a_j\}$, where j is the total number of attributes. Different types of auctions are categorized based on some predefined attributes from the vast feature space of online auctions. The feature space may include average bid amount, average bid rate, number of bids, item type, seller reputation, opening bid, closing bid, quantity available, type of auction, duration of the auction, buyer reputation and many more. In this paper, to classify different types of auctions, we focus on only a set of attributes; opening bid,

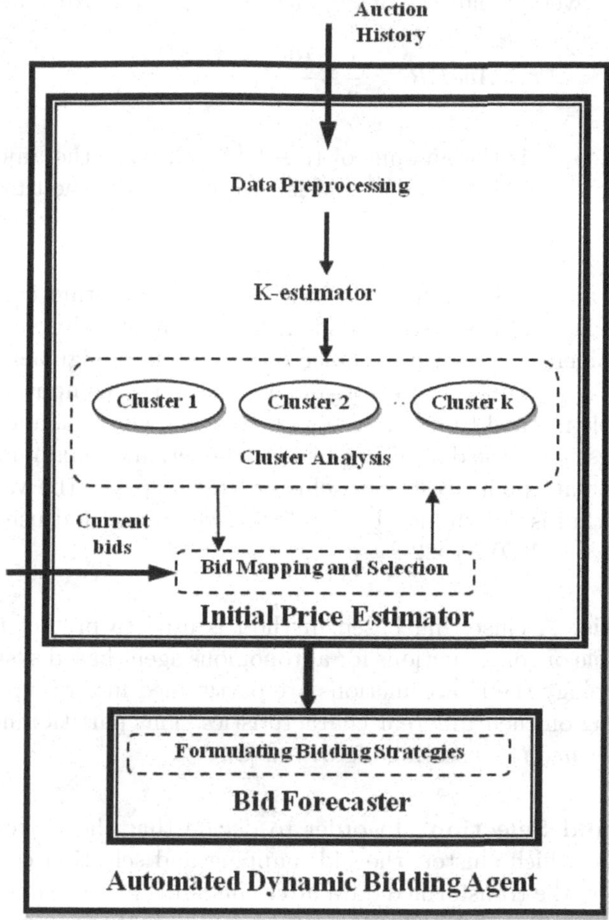

Fig. 1. Automated Dynamic Bidding Agent

closing price, number of bids, average bid amount and average bid rate. Now $A = \{OpenB_i, CloseP_i, NUM_i, AvgB_i, AvgBR_i\}$

where A be the set of attributes for an auction

$OpenB_i$ be the starting price of ith auction

$CloseP_i$ be the end price of ith auction

NUM_i be the total number of bids placed in i^{th} auction

$AvgB_i$ be the average bid amount of ith auction and can be calculated as $Avg(B_1, B_2, ...B_l)$ where B_1 is the 1st bid amount, B_2 be the second bid amount and B_l is the last bid amount for i^{th} auction.

$AvgBR_i$ be the average bid rate of ith auction and can be calculated as

$$AvgBR_i = \frac{1}{n}\Sigma\frac{B_{i+1} - B_i}{t_{i+1} - t_i} \tag{1}$$

where $1 \leq i \leq n$, B_{i+1} is the amount of $(i+1)^{th}$ bid, B_i is the amount of ith bid, t_{i+1} is the time at which $(i+1)^{th}$ bid is placed and t_i is the time at which i^{th} bid is placed.

K-estimator. To decide the value of k in $k - means$ algorithm is a recurrent problem in clustering and is a distinct issue from the process of actually solving the clustering problem. The optimal choice of k is often ambiguous, increasing the value of k always reduce the error and increases the computation speed. The most favorable method to find k adopts a strategy which balances between maximum compression of the data using a single cluster, and maximum accuracy by assigning each data point to its own cluster. In this paper the value of k in $k - means$ algorithm is determined by employing elbow method using one way analysis of variance (ANOVA) [22].

Cluster Analysis. A clustering based method is used to predict the closing price of the multiple ongoing auctions for autonomous agent based system. In the proposed methodology the input auctions are partitioned into groups of similar auctions depending on their different characteristics. This partitioning has been done by using $k - means$ clustering algorithm [22].

Bid Mapping and Selection. In order to decide that the current ongoing auctions belong to which cluster, the bid mapping and selection component is activated. Based on the transformed data after clustering and the characteristics of the current auctions, it nominates the cluster for each of the ongoing auction to select the auction to participate in and to predict its closing price.

We observe that in 92 of the 149 auctions of our dataset, winner first appear in the last hour of the auction, which account for the 62% of the total auctions, consistent with the late bidding attitude of the bidders recognized in the online auction literature [10] [11] [12]. Clustering has divided the auction data into the groups of auctions having distinct range of average bid rate ($AvgBR$) values. It has been observed that the value of $AvgBR$ in 78% of the completed auctions belong to the same cluster as at the beginning of the last hour of the auction. So the ongoing auctions are mapped to the clusters based on their $AvgBR$ value in the beginning of the last hour. The methodology for selecting the auction for participation and predicting the closing price is described as follows:

Let OA be the set of multiple ongoing auctions and $OA = OA_1 \cup OA_2 \cup ... \cup OA_k$. where OA_i be the set of ongoing auctions belonging to the ith cluster and $i = 1, 2, ...k$ where k is the total number of clusters.

Then $OA_i = \{OA_{il}, OA_{i2}, ...OA_{in}\}$ where n is the total number of ongoing auctions belonging to the ith cluster.

Let $AvgB_i$ be the set of average bid amounts of the ongoing auctions belonging to the ith cluster at the beginning of the last hour then $AvgB_i = \{AvgB_{il}, AvgB_{i2}, ...AvgB_{in}\}$.

Let $AvgBC_i$ be the average bid amount of all the auctions in the ith cluster. A subset of OA will be selected where $AvgB_{in} < AvgBC_i$, where $i = 1, 2, ...k$. Then the auction with the minimum value of the average bid amount will be selected for the participation.

The closing price prediction task for the selected auction is treated as a multiple regression task [22]. The predicted closing price of the selected auction is treated as the initial price (p_i) by the bid forecaster.

2.2 Bid Forecaster

This section forecasts the bid amount for the selected ongoing auction by designing the bidding strategies for different bidding behaviors of the buyer. We classify the late bidders' behavior based on the single or multiple bids placements and are further categorized based on the time of the bid placement Fig. 2.

The late bidders who place single bid are Mystical and Sturdy. The mystical bidders place single bid in the last five minutes of the auction. The sturdy bidders place single bid in the last hour of the auction. The late bidders who place multiple bids are Desperate and Strategic. The desperate bidders continuously bid for an item multiple numbers of times without any other participants in between. The strategic bidders also place multiple bids to win the auction in the last hour; however, these bidders increment their bid amount strategically based on the bids placed by the other participants.

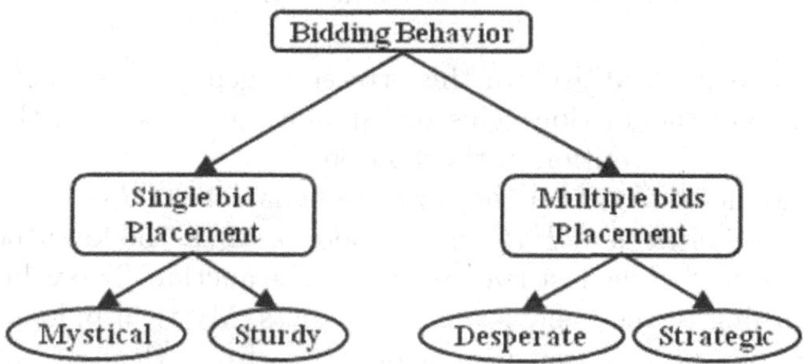

Fig. 2. Bidding Behaviors of the late bidders

The bidding strategies are designed for the above bidding behaviors of the buyers. The bidding agent models the bid amount based on the initial price (p_i) and the negotiation decision functions by Faratin et al [21] for each bidding bahevior in Fig. 2. The purpose of these functions is to determine how to compute the bid amount at a particular moment of time. These negotiation decisions may depend upon the remaining time of the auction or on the bids placed by the other participants (competition).

Let $F(t)$ be the function to determine the bid amount based on the remaining time left and let $F_c(t)$ be the function to determine the bid amount based on the competition in the auction. Let the agent bids at time $0 \leq t \leq t_{max}$. Agent′s bidding limit is $[min_b, max_b]$. Then,

$$F(t) = min_b + \alpha(t)(max_b - min_b) \tag{2}$$

where $\alpha(t) = k + (1-k)(\frac{min(t,t_{max})}{t_{max}})^{\frac{1}{\beta}}$

A wide range of time dependent functions can be calculated by varying the value of $\alpha(t)$.

where $0 \leq \alpha(0) \leq 1$, $\alpha(0) = k$ and $\alpha(t_{max}) = 1$

k is a constant and $0 \leq k \leq 1$. When k is multiplied by the size of the bid interval, it gives the value of the starting bid amount. β is a constant which belongs to R^+. A number of possible bidding regulations can be obtained by varying the value of β. When $\beta < 1$, minimum bid amount will be maintained until the t_{max} is almost reached and when $\beta > 1$, agent quickly goes to its reservation price p_r (maximum willingness to pay) where $p_r = p_i = max_b$.

$F_c(t)$ compute the bid amount at time t based on the attitude of the other participants who placed the previous bids. To calculate $F_c(t)$ at a particular moment of time t, agent reproduces the behavior of the other participants $\delta \geq 1$ steps before in percentage terms. Where $n > 2\delta$.

$$F_c(t_{n+1}) = min(max(\frac{F'(t_{n-2\delta+2})F(t_{n-1})}{F'(t_{n-2\delta})}, min_b), max_b) \tag{3}$$

Where $F'(t)$ be the bids placed by the other participants at time t. The value of min_b depends on the bidding strategy followed.

Mystical Bidding Strategy. In this strategy agent places single bid in the last five minutes of the auction. This bid amount depends upon the remaining time as well as the competition in the auction.

The bid amount at time t for the mystical behavior will be calculated as the average of $F(t)$ as in (2) and $F_c(t)$ in (3). Here, min_b is the lower bound of the bid value at the start of the last five minutes of the auction. The values for k and β are set according to the behavior of the bidders. Mystical bidders commonly possess two behaviors, first, they may be desperate to get the item, and secondly, they may be willing to bargain for that item. For the mystical bidders having desperate behavior, the value of k will be high and $\beta > 1$, since this type of bidders bid at a price near to the reservation price p_r. On the other hand, for the mystical bidders with a desire for bargain behavior, the value of k will be low and $\beta < 1$.

Sturdy Bidding Strategy. This strategy is similar to the mystical bidding behavior with an exception of the time of placing a bid. A bidder with sturdy behavior places a single bid at the beginning of the last hour of the auction based on the remaining time and the competition in the auction.

$F(t)$ and $F_c(t)$ functions similar to the mystical behavior are used to compute the bid amount, but here min_b is the lower bound of the bid amount at the beginning of the last hour. The sturdy bidders also appeared in two behaviors; desperate, and desire to get bargain. The values for k and β for sturdy bidders with desperate and desire to get bargain behaviors follow the same conventions as in the mystical bidders.

Desperate Bidding Strategy. Desperate bidders place multiple bids continuously during the last hour of the auction. The first bid placed by the bidder depends on the remaining time left and the competition in the auction. Rest of the bids he places based on the remaining time left of the auction.

The starting bid will be calculated as the average of $F(t)$ and $F_c(t)$. All the other bids will be calculated from $F(t)$. The value of the min_b is the same as in the sturdy bidding strategy. As a desperate bidder starts bidding at a value close to his valuation for the item, the values for k are high for this bidding strategy. Here $\beta > 1$, since desperate bidders tend to quickly reach at p_r before the deadline is reached by placing multiple bids continuously.

Strategic Bidding Strategy. Strategic bidding strategy is similar to the desperate behavior with one key difference in the way of placement of the bids. Strategic bidders place each and every bid strategically based on the bids placed by the other participants in the auction. They continue bidding till the bid amount reaches their reservation price p_r.

Each bid will be calculated as the average of $F(t)$ and $F_c(t)$. The value of the min_b is the same as in the sturdy bidding strategy. Strategic bidders do not start biding at an amount close to p_r; rather they increase their bid amount slowly based on the other bids in the auction. So the values for k are low and $\beta < 1$ for the strategic behavior.

3 Experimentation

The performance of our bidding agent is assessed by undertaking an empirical evaluation of the automated dynamic bidding agent (ADBA) in two phases. In the first phase, the methodology for the initial estimation is validated, and in the second phase, the success rate of the bidding agent following different behaviors of the buyers is analyzed. The dataset for our experimentation includes the complete bidding records for 149 auctions for new Palm M515 PDAs. The statistical description of the data is explained in [23].

3.1 Initial Price Estimation

The initial price estimator selects an auction to participate and estimates its closing price based on the clustering and bid mapping approach. This predicted closing price act as the initial price for the bid forecaster. Auction for participation is selected based on the bid mapping and selecting technique. Closing price of the selected auction is predicted by exploiting clustering and multiple linear regression approach [22]. In the proposed approach closing price of an online auction is predicted in two scenarios. First, it is predicted by exploiting multiple linear regressions on whole input auctions data. Second, by exploiting multiple linear regression on each cluster which are generated by applying $k - means$ algorithm on whole input auctions data. The results are evaluated by comparing the root mean square errors (RMSEs) in both of these scenarios. Experimental results demonstrated less RMSEs for the prediction results when multiple linear regression is applied on each cluster rather than on whole input auctions data [22].

3.2 Success Rate of the Bidding Agents

Bid forecaster forecasts the bid amount for the selected ongoing auction based on the different bidding strategies followed by the bidder. A bidding agent with the bidding behavior which only depends on the $F(t)$ not on $F'(t)$ is opted as a basis for the comparison. To compute the function $F_c(t_{n+1})$, the initial values of $F'(t_{n-2\delta+2})$, $F'(t_{n-2\delta})$ and $F(t_{n-1})$ are calculated at $\delta = 1$ for all the bidding strategies i.e. Mystical, Sturdy, Desperate and Strategic. For sturdy, desperate and strategic behaviors $\delta = 1$ at the beginning of the last hour of the auction. For Mystical behavior, $\delta = 1$ at the beginning of the last five minutes of the auction. $F'(t_{n-2\delta+2})$ and $F'(t_{n-2\delta})$ are calculated as the percentage of the reservation price (p_r) reached at time $t = n - 2\delta$ and at $t = n - 2\delta + 2$. $F(t_{n-1})$ is the average of $F'(t_{n-2\delta+2})$ and $F'(t_{n-2\delta})$. The percentage of the reservation price (p_r) reached at time $t_{n-2\delta+2}$ and $t_{n-2\delta}$ for all the bidding strategies are as shown in the Table 1 ($\delta = 1$).

Table 1. Percentage of reservation price (p_r) at time $t = n - 2\delta + 2$ and $t = n - 2\delta$ for different bidding strategies

Bidding Strategy	%age of p_r at $t_{n-2\delta+2}$	%age of p_r at $t_{n-2\delta}$
Mystical	97	95
Sturdy	88	86
Desperate	94	91
Strategic	88	87

In this set of experiments, the values for k and β depend on the selected bidding strategy for the bidding agent. Let MST_d and MST_b represent the mystical bidders with desperate and desire to bargain behavior respectively. STD_d and STD_b represent the sturdy bidders with desperate and desire to bargain behavior respectively. DSP and STG represent the desperate and strategic bidders. The values for k and β chosen for these behaviors are shown in Table 2.

The bid amount selected by these bidding strategies depend upon the remaining time of the auction and on the bids placed by the other participants (competition). There are basically two types of bidders. They may be desperate to have the item or they may be willing to bargain to acquire the item being auctioned. Desperate bidders start bidding at higher price close to his reservation value p_r and their bid amount is less affected by the amount of bids placed

Table 2. Choice of k and β for different bidding strategies

Bidding Strategy	k	β
MST_d	$0.6 \leq k \leq 1$	$\beta > 1$
MST_b	$0 \leq k \leq 0.3$	$\beta < 1$
STD_d	$0.6 \leq k \leq 1$	$\beta > 1$
STD_b	$0 \leq k \leq 0.3$	$\beta < 1$
DSP	$0.6 \leq k \leq 1$	$\beta > 1$
STG	$0 \leq k \leq 0.6$	$\beta < 1$

by the other bidders due to their desperate behavior. MST_d, STD_d,and DSP bidding strategies follow this type of behavior. The bidders who are willing to bargain always bids strategically based on the bids placed by the other competitors. MST_b , STD_b and STG follow this type of behavior. As the bidding strategies described above select the bids based on the remaining time as well as on the bids placed by the other participants (competition), these strategies will be successful when the bidder has a desire to bargain behavior. So we will discuss the performance of the agents acting strategically based on the bids placed by the other participants. The bidding agents with the behavior which only depends on the $F(t)$ not on $F'(t)$ are opted as the basis for the comparison for each bidding strategy. The performance of all the agents following these bidding strategies is shown in Fig. 3. The performance of the agents is measured in terms of the number of times he wins the auction.

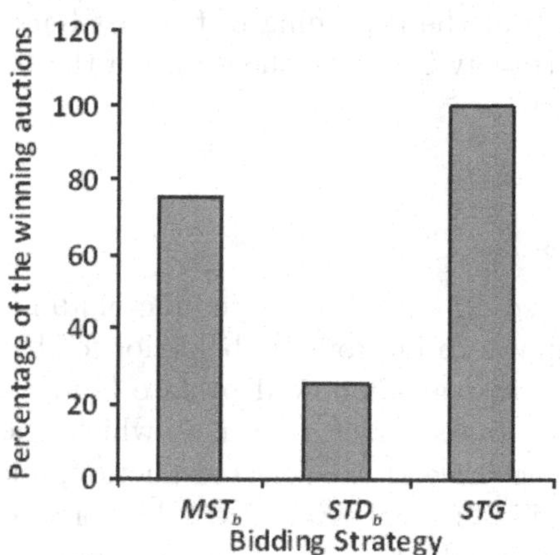

Fig. 3. Percentage of winning auction comparision

Results show that the strategy STG performed best followed by MST_b and then STD_b. The agent following strategy STG has won all the auctions because he bids strategically based on the amount of the bids placed by the other participants and the remaining time of the auction. He continue bidding till he gets the

Bidding strategy	Desire for bargain		
	High	Medium	Low
STG	✓	✓	✓
MST_b	✓	✓	X
STD_b	✓	X	X

Fig. 4. Choosing Bidding strategy

item or the bid reaches at his reservation price. (In our experiments we assume that closing price of the item $\leq p_r$ for the agent and p_r for the agent $\geq p_r$ for the competitor).

To explain the performance of the agents following MST_b and STD_b strategy, the levels of the desire for bargain attitude are distinguished as low, medium and high depending on the value of k . It has been observed that the mystical agent wins in 75% of the auctions. The winning bidders have medium to high level of desire for bargain for having the item. The bidders with low level of desire for bargain place bids based on the remaining time of the auction. Their bid amount has been less affected by the bids of the competitors because of their attitude of single bidding in last five minutes of the auction. The agent following STD_b wins in 25% of the auctions. The winning bidders possess high level of desire for bargain behavior. The sturdy bidders with low and medium level of desire for bargain place bids based on the remaining time of the auction. Their bid amount has been less affected by the bids of the competitors because of their attitude of single bidding at the beginning of the last hour of the auction. The most suitable bidding strategy based on the desire of the bargain for the bidder has been shown in Fig. 4.

4 Conclusions

In this paper we presented an automated dynamic bidding agent (ADBA) that uses data mining techniques to improve its behavior for bid forecasting according to different bidding strategies followed by late bidders. The bidding agent primarily performs three tasks; first, it decides which auction to bid in, secondly, predicts the closing price of the selected auction, and finally, models the bid amount for a given bidding behavior of the bidders. Auction selection and the closing price prediction adopt a clustering, and bid mapping and selection approach. The ADBA forecasts the bid amount for the selected ongoing auction by designing the bidding strategies based on the bidding behavior of the bidders. Bidding strategies are designed according to the predicted closing price and the negotiation decision functions for the specific bidding behaviors. By allowing negotiation decision functions for different bidding behaviors of the buyers, ADBA presents improved bid forecasting results than the DC mechanism [15]. The outcome of the clustering based model for the closing price prediction is compared with the classic model for price prediction. The improvement in the error measure for each cluster for a set of attributes gives support in favor of the proposed model using clustering. Our experimental results demonstrated the proficiency of the designed bidding strategies for the late bidders with different levels of desire for bargain behavior.

References

1. Ockenfels, A., Reiley Jr., D.H., Sadrieh, A.: Online auctions. National Bureau of Economic Research Cambridge, Mass, USA (2006)
2. Haruvy, E.: Internet auctions. Foundations and Trends in Marketing 4(1), 1–75 (2009)
3. Park, Y.H., Bradlow, E.T.: An integrated model for bidding behavior in Internet auctions: Whether, who, when, and how much. Journal of Marketing Research 42(4), 470–482 (2005)
4. Jank, W., Zhang, S.: An automated and data-driven bidding strategy for online auctions. Journal of Computing 23(2), 238–253 (2011)
5. Anthony, P., et al.: Autonomous agents for participating in multiple online auctions. In: IJCAI Workshop on E-Business and the Intelligent Web, Seattle, USA, pp. 54–64 (2001)
6. Greenwald, A., Stone, P.: Autonomous bidding agents in the trading agent competition. IEEE Internet Computing 5(2), 52–60 (2001)
7. Bapna, R., et al.: User heterogeneity and its impact on electronic auction market design. An Empirical Exploration. Mis Quarterly 28(1), 21–43 (2004)
8. Shah, H., et al.: Mining eBay: Bidding strategies and shill detection. In: WE-BKDD 2002-MiningWeb Data for Discovering Usage Patterns and Profiles, pp. 17–34 (2003)
9. Trevathan, J., Read, W.: Detecting shill bidding in online English auctions. In: Handbook of Research on Social and Organizational Liabilities in Information Security (2008)
10. Du, L., Chen, Q., Bian, N.: An Empirical Analysis of Bidding Behavior in Simultaneous Ascending-Bid Auctions. In: International Conference on E-Business and E-Government (ICEE). IEEE (2010)
11. Rasmusen, E.B.: Strategic implications of uncertainty over one's own private value in auctions. The BE Journal of Theoretical Economics 6, 7 (2006)
12. Ockenfels, A., Roth, A.E.: Late and multiple bidding in second price Internet auctions: Theory and evidence concerning different rules for ending an auction. Games and Economic Behavior 55(2), 297–320 (2006)
13. Xuefeng, L., et al.: Predicting the final prices of online auction items. Expert Systems with Applications 31(3), 542–550 (2006)
14. Nikolaidou, V., Mitkas, P.: A Sequence Mining Method to Predict the Bidding Strategy of Trading Agents. In: Agents and Data Mining Interaction, pp. 139–151 (2009)
15. Kehagias, D.D., Mitkas, P.A.: Efficient E-Commerce Agent Design Based on Clustering eBay Data. In: International Conferences on Web Intelligence and Intelligent Agent Technology Workshops. IEEE/WIC/ACM (2007)
16. Heijst, D., Potharst, R., Wezel, M.: A support system for predicting ebay end prices. Econometric Institute Report (2006)
17. Ghani, R., Simmons, H.: Predicting the end-price of online auctions (2004)

18. Zhang, S., Jank, W., Shmuel, G.: Real-time forecasting of online auctions via functional k-nearest neighbors. International Journal of Forecasting 26(4), 666–683 (2010)
19. Cao, L., Weiss, G., Yu, P.S.: A Brief Introduction to Agent Mining. Journal of Autonomous Agents and Multi-Agent Systems 25, 419–424 (2012)
20. Cao, L., Gorodetsky, V., Mitkas, P.A.: Agent Mining: The Synergy of Agents and Data Mining. IEEE Intelligent Systems 24(3), 64–72 (2009)
21. Faratin, P., Sierra, C., Jennings, N.R.: Negotiation decision functions for autonomous agents. Robotics and Autonomous Systems 24, 159–182 (1998)
22. Kaur, P., Goyal, M., Lu, J.: Data mining driven agents for predicting online auction's end price. In: IEEE Symposium on Computational Intelligence and Data Mining (CIDM), pp. 141–147. IEEE, Paris (2011)
23. Kaur, P., Goyal, M., Lu, J.: Pricing Analysis in Online Auctions Using Clustering and Regression Tree Approach. In: Cao, L., Bazzan, A.L.C., Symeonidis, A.L., Gorodetsky, V.I., Weiss, G., Yu, P.S. (eds.) ADMI 2011. LNCS, vol. 7103, pp. 248–257. Springer, Heidelberg (2012)

Trading Strategy Based Portfolio Selection for Actionable Trading Agents

Wei Cao, Cheng Wang, and Longbing Cao

Advanced Analytics Institute,
University of Technology, Sydney
Wei.Cao@student.uts.edu.au,
cescwang@gmail.com,
LongBing.Cao@uts.edu.au

Abstract. Trading agents are very useful for supporting investors in making decisions in financial markets, but the existing trading agent research focuses on simulation on artificial data. This leads to limitations in its usefulness. As for investors, how trading agents help them manipulate their assets according to their risk appetite and thus obtain a higher return is a big issue. Portfolio optimization is an approach used by many researchers to resolve this issue, but the focus is mainly on developing more accurate mathematical estimation methods, and overlooks an important factor: trading strategy. Since the global financial crisis added uncertainty to financial markets, there is an increasing demand for trading agents to be more active in providing trading strategies that will better capture trading opportunities. In this paper, we propose a new approach, namely trading strategy based portfolio selection, by which trading agents combine assets and their corresponding trading strategies to construct new portfolios, following which, trading agents can help investors to obtain the optimal weights for their portfolios according to their risk appetite. We use historical data to test our approach, the results show that it can help investors make more profit according to their risk tolerance by selecting the best portfolio in real financial markets.

Keywords: Trading Strategy, Portfolio Selection, Trading Agent.

1 Introduction

Trading in electronic markets is increasingly recognized as a promising domain for agent technology within the artificial intelligence field. Trading agents play a significant role in evaluating programmed trading techniques and developing automated strategies in electronic financial markets [15,17], and consequently, the development of trading agent competition [4] (for example, design tradeoffs [12]) has received increasing attention.

Most of the current trading agent research focuses on problems designed in an artificial marketplace with simulated data [3,14]; little attention has been paid to problems in real financial markets, which leads to the lack of practical application for business people, who have their own business preferences. How

L. Cao et al.: ADMI 2012, LNAI 7607, pp. 191–202, 2013.

trading agents help investors to manipulate their assets according to their risk appetite and then obtain greater return is a big issue.

Many researchers and market participants may choose the right portfolio to resolve the above issue. Because the application of a right portfolio is a very powerful tool in risk control [11], that is to say investors can mitigate risk by allocating capital to various assets with different weights. Portfolio theory was first introduced by Harry Markowitz in [5] and it has attracted significant attention in practice and in the academic field. The principal rule is to maximize the expected return for a given level of risk, or to minimize the risk based on expected return. With the continuous effort of various researchers, Markowitz's work has been widely extended, but most of the literatures focuses on how to find a better mathematic method to estimate corresponding parameters and gain an optimal return which is closer to the theoretical optimal return. Trading strategy as an important factor that is often ignored in the literature to date.

Trading strategies have been proposed in financial literatures and trading houses to support trading investment decisions. The right trading strategies can assist trading agents in determining the right actions at the right time at the right price [3], so if we import trading strategy into the portfolio optimization problem, we can capture more profitable trading opportunities. Secondly, in practical financial markets, investors always use trading strategies to conduct transactions, so it is not reasonable if we ignore the trading strategies that have been used in the portfolio selection process. In addition, from the view of trading strategy independently, different trading strategies will generate different signals (buy, sell, hold) for one asset, which will lead to different returns. It is not available to choose the profitable one for every investor for their different risk appetite. A trading strategy which produce more return at a period may mean more volatility, namely a higher risk is accompanied when compared to other trading strategies. From this point, the demand for trading agents to be more actionable to provide more proper trading strategies is increasing.

From the above we can see that real financial markets are full of uncertainty, it is not enough to select right portfolios to obtain more profit only through improved mathematical methods. Trading strategy as an important factor also need to be accounted. As for trading strategies, since the global financial crisis in 2007 adds the uncertainty in financial markets, there is increasing demand for trading agents to be more actionable to provide trading strategies for better trading opportunities in financial markets. All of these lead to our concern on how to combine trading strategies and portfolio selection process. In this paper, we develop a new approach to select the right portfolio for trading agents, namely trading strategy based portfolio selection. The system works as follows: for given trading strategies and assets at a time window, firstly we combine every asset and every trading strategy, then we compare the performance of corresponding combinations for each asset; after this we delete the combinations with bad performance. In addition, we use an improved estimation method to obtain the optimal weights for the selected combinations.

The rest of the paper is organized as follows. Section 2 reviews the work related to this paper: trading strategy, portfolio theory and related estimation methods. The framework of our trading agent is illustrated in Section 3. Empirical experiments and evaluation are illustrated in Section 4. We draw conclusions in Section 5.

2 Background

2.1 Trading Strategy

Trading strategies are widely used by financial market traders to assist them in determining their investment or speculative decisions [2], in which way they can make higher profit with lesser risk. A trading strategy indicates when a trading agent can take what trading actions under certain market situation. It is governed by a set of rules that do not deviate. Herein, an example is given: a general Moving Average (MA) based trading strategy.

Example (MA trading strategy) An MA trading strategy is simply an average of current and past prices over a specified period of time. An MA of length l at time t is calculated as

$$M_t(l) = \frac{1}{l}\sum_{i=0}^{l-1} P_{t-i} \tag{1}$$

where p_{t-i} is the price at time $t-i$. Many kinds of trading strategies can be formulated based on MA. A filtered MA strategy (denoted by MA(l,θ)) compares the current price p_t to its MA value M_t, if p_t rises above M_t by more than a certain percent θ, the security is bought and held until the price falls below MA more than θ percent at which the security is sold. Such trading signals s_t are generated according to Equation (2).

$$s_t = \begin{cases} -1 & \begin{array}{l} if\, P_t > (1+\theta)M_t(l)\, and\, P_{t-u} < M_{t-u}(l) \\ and\, M_{t-i}(l) \le P_{t-i} \le (1+\theta)M_{t-i}(l), \\ \forall i \in \{1,\cdots,u-1\} \end{array} \\ 1 & \begin{array}{l} if\, P_t < (1-\theta)M_t(l)\, and\, P_{t-v} > M_{t-v}(l) \\ and\, M_{t-i}(l) \ge P_{t-i} \ge (1-\theta)M_{t-i}(l), \\ \forall i \in \{1,\cdots,v-1\} \end{array} \\ 0 & otherwise \end{cases} \tag{2}$$

where u and v are arbitrary positive integers, -1 means 'buy', 1 means 'sell' and 0 means 'hold' or 'no action'. In real life trading, trading strategies can be categorized into different classes and a trading strategy class can further be instantiated into different types of trading strategies by different constraint filters. For instance, Moving Average MA discussed above is a common trading strategy, it can be instantiated into several types by different applications of θ. If

$\theta = 0$, it is shrinks to a simple MA. Investors can obtain various results through changing parameters l and θ according to their own risk appetite.

Another three classes of trading strategies: Filter Rules, Support and Resistance and Stochastic Oscillator strategies [13,16] are applied in the experiments in Section 4 which are described below briefly.

Filter Rules (denoted by $FR(l, \theta)$) indicates that if the daily closing price of a particular security moves up at least θ percent, buy and hold until its price moves down at least θ percent from a subsequent high, at which time sell and go short, the short position is maintained until the daily closing price rises at least θ percent above a subsequent low at which time one covers and buys. A high/low can be defined as the most recent closing price that is greater/less than the l previous closing prices.

Support and Resistance (denoted by $SR(l)$) suggests to buy/sell when the closing price exceeds the maximum/minimum price over the previous l days.

Stochastic Oscillator strategies (denoted by $K/D(x, y)$) generate a buy/sell signal when $\%K$ line crosses a $\%D$ line in an upward/downward direction. x, y here are the numbers of periods used to compute the $\%K$ and $\%D$.

2.2 Portfolio Theory

Every investor who wants to get more return subject to a given risk through allocating his capital among different assets within a market will face asset portfolio problem. Markowitz is the father of portfolio theory, he introduced the theory in [5] and then won the Nobel Prize in Economic Sciences for the theory. According to this theory, investors respond to the uncertainty of the market by minimizing risk subject to a given level of expected return, or equivalently maximizing expected return for a given level of risk. These two principles led to the formulation of an efficient frontier from which an investor could choose his or her preferred portfolios, depending on individual risk tolerance. This is done by choosing the quantities of various assets cautiously and taking mainly into consideration the way in which the price of each asset changes in comparison to that of every other asset in the portfolio. In other words, the theory uses mathematical models to construct an ideal portfolio for an investor that gives maximum return depending on his or her risk appetite by taking into consideration the relationship between risk and return.

The mean-variance (MV) model introduced by Markowitz plays an important role in the portfolio problem. It is a bi-criteria optimization problem in which the expectation of return and risk are combined. It has triggered a large amount of research activities in the field of finance for optimal portfolio choice, but it is fraught with practical problems. The main problem is that the estimation error exists in computing the expected returns and the covariance matrix of asset returns. So far researchers (for example, [8,1]) all believe that the reason of the estimation error is the "optimal" return is formed by a combination of returns from an extremely large number of assets, this will lead to over-prediction. In recent years, many attempts have been undertaken to ease the amplification of the estimation error. In [9], authors suggested that imposing the nonnegativity constraint on portfolio weights can make the process more intuitive. Authors in [1] developed new bootstrap-corrected estimations for the optimal return and its asset allocation and proved these estimates are proportionally consistent with their theoretic counterpart.

In our paper, we use a new and improved estimation method introduced by [10] to compute our trading strategies based portfolios. Authors in [11] pointed out and analyzed the limitation of estimators proposed by [1]. Based on this, they presented a new and improved estimator for the optimal portfolio return which can circumvent the limitation. They proved that the new and improved estimated return is consistent when sample size $n \to \infty$ and the dimension to sample size ratio $c/n \to z \in (0, 1)$. They also illustrated that their estimators dramatically outperformed than traditional estimators through their simulation, especially when c/n is close to 1. This is very useful for our issue, because when combined different trading strategies to assets, the sample size c will become much larger than the number in the situation only consider allocation of assets.

3 Modeling Framework

3.1 Trading Agent Framework

The working mechanism of trading agents is represented in Fig.1.

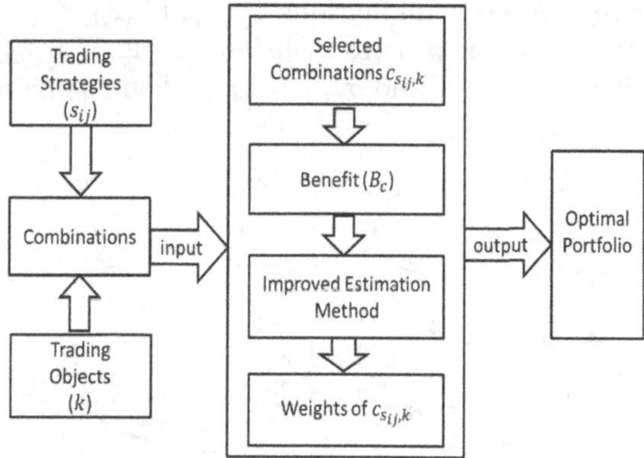

Fig. 1. Framework of actionable trading agents

The working mechanism of actionable trading agents is as follows:

Step 1. For given trading strategies, the agents combine every trading object with every trading strategy. A combination $C_{s_{ij},k}$ indicates the state that an trading agent uses the $s_{ij}th$ trading strategy on kth trading object. s_{ij} represents the jth trading strategy taken from trading strategy class i. Fig.2 is an example of the combinations of two trading objects and four trading strategies from two trading strategy classes. The combinations are inputs in the process.

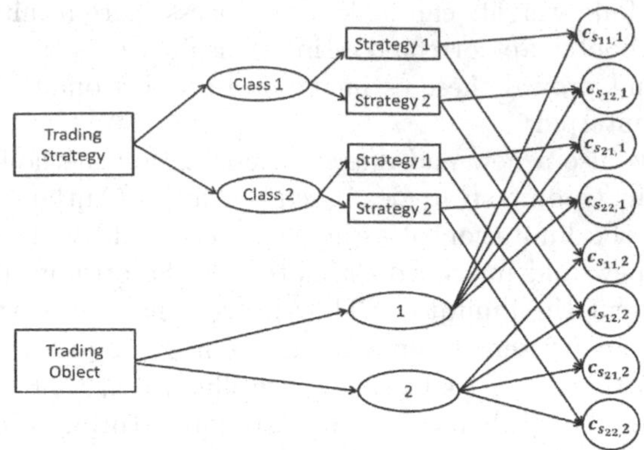

Fig. 2. An example of combination

Step 2. In this step, we do a selection of the combinations obtained from Step 1. For each trading object, we delete those corresponding combinations with bad performance, and leave the remaining for trading agents. Here we use two variables to test the applicability of the obtained combinations, namely return R and volatility σ. For a trading object, we need to compute the return and volatility with the corresponding combinations; then we compared each combination with others to decide which combinations can be abandoned.

For example, there are four combinations $C_{s_{11},1}$, $C_{s_{12},1}$, $C_{s_{21},1}$ and $C_{s_{22},1}$ for trading object 1, the selection process is listed in Fig.3. $\exists C_{m,1}, C_{n,1},\ m, n \in \{s_{11}, s_{12}, s_{21}, s_{22}\}$, if $R_{m,1} < R_{n,1}$ and $\sigma_{m,1} > \sigma_{n,1}$, then $C_{m,1}$ should be deleted.

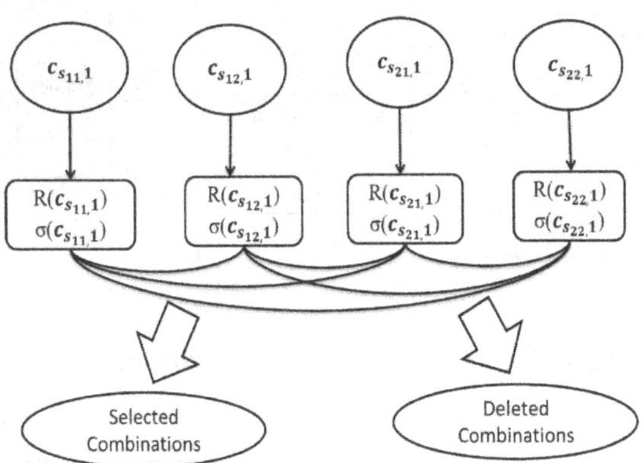

Fig. 3. An example of selection

Step 3. We calculate the benefit of the combinations obtained in Step 2. Here we use the log return to generate the benefit B_C according to signals produced by the corresponding trading strategies. A trading strategy s_{ij} may produce three trading signals in transactions: $\{-1, 0, 1\}$, where -1 indicates a buy signal, 1 indicates a sell signal and 0 means hold or no action. In our process, for example, the benefit for a combination at time t is computed as $B_{C_t} = \log p_t - \log p_{t-1}$ from the generation of a buy signal to that of a sell signal. The benefit for the trading object is 0 in a interval from a sell signal is generated to a buy sinal is generated. Because at this period the object was sold, so the money did not have any change. In addition, in order to simplify the complexity in calculation and make it easier to compare, the volume here equals to 1.

Step 4. Based on the benefit obtained from Step 3, in this step we put these benefits into an improved estimation method to calculate the optimal weights for the selected combinations with a given level of risk. A brief introduction of the method was put in Section 3.2. According to the optimal weights, the trading agent can obtain the optimal portfolio which lead to optimal return.

3.2 Improved Estimation on Markowitz Mean-Variance Analysis

Suppose that there exist c assets in a portfolio, $p = (p_1, p_2, ..., p_c)^T$, and the return denoted by $B = (b_1, b_2, ..., b_c)^T$ follows a multivariate normal distribution with mean $\mu = (\mu_1, \mu_2, ..., \mu_c)^T$ and covariance matrix Σ. According to Markowitz's mean-variance model, an investor who invests capital M on the assets wants to maximize her/his return subject to a given risk, namely try to find an optimal weight $W = (w_1, w_2, ..., w_c)^T$ to maximize portfolio return while keeping the risk level under a specific level σ_0^2. The above maximization problem can be formulated below:

$$R = max\, w^T \mu \;\, subject\ to\ w^T \Sigma w \leq \sigma_0^2 \ and\ w^T \mathbf{1} \leq 1 \tag{3}$$

where $\mathbf{1}$ represents a c dimensional vector of ones and σ_0^2 indicates the risk level that an investor can suffer. Here we assume that the total portfolio weight $\sum_{i=1}^{c} w_i \leq 1$ for without of generality. Hence, the goal is to find R and W that satisfied the above equation.

From [6,7,1] we can obtain the analytical solution to the problem stated in Equation (3) as follows:

$$w = \begin{cases} \frac{\sigma_0}{\sqrt{\mu^T \Sigma^{-1} \mu}} \Sigma^{-1} \mu, & if \frac{\mathbf{1}^T \Sigma^{-1} \mu}{\sqrt{\mu^T \Sigma^{-1} \mu}} \sigma_0 < 1 \\ \\ \dfrac{\Sigma^{-1} \mathbf{1}}{\mathbf{1}^T \Sigma^{-1} \mathbf{1}} + d[\Sigma^{-1} \mu - \dfrac{\mathbf{1}^T \Sigma^{-1} \mu}{\mathbf{1}^T \Sigma^{-1} \mathbf{1}} \Sigma^{-1} \mathbf{1}], & otherwise \end{cases} \tag{4}$$

where $d = \sqrt{\frac{(\mathbf{1}^T \Sigma^{-1} \mathbf{1}) \sigma_0^2 - 1}{(\mu^T \Sigma^{-1} \mu)(\mathbf{1}^T \Sigma^{-1} \mathbf{1}) - (\mathbf{1}^T \Sigma^{-1} \mu)^2}}$

Equation (4) provides the solution, and the critical issue here is how to estimate μ and Σ because they are unknown in practical applications. A simple and natural way is to use the corresponding sample mean $\bar{X}$ and sample covariance matrix S, respectively. This type of estimation has been proved to be a very poor estimate method in [1] for its over-prediction problem, namely the expected return using this method is always larger than the theoretic optimal return.

Instead of using the above sample estimates, in this paper we use another estimate method developed by [10], they propose the unbiased estimators to construct the efficient estimators of the optimal return. The expectation properties of these estimators are as follows:

$$\begin{cases} E[\frac{n-c-2}{n-1}S^{-1}] = \Sigma^{-1} \\ E[\frac{n-c-2}{n-1}\mathbf{1}^T S^{-1}\bar{X}] = \mathbf{1}^T \Sigma^{-1}\mu \\ E[\frac{n-c-2}{n-1}\mathbf{1}^T S^{-1}\mathbf{1}] = \mathbf{1}^T \Sigma^{-1}\mathbf{1} \\ E[\frac{n-c-2}{n-1}\bar{X}^T S^{-1}\bar{X} - \frac{c}{n}] = \mu^T \Sigma^{-1}\mu \end{cases} \tag{5}$$

where n is the number of sample from $N_c(\mu, \Sigma)$ distribution, $\bar{X}$ and S are sample mean and sample covariance matrix, respectively. We assume $n - c - 2 > 0$ here.

The above unbiased estimation can be used to correct the corresponding return and weight in the optimization issue, and the improved estimators can be obtained as follows:

$$w_I = \begin{cases} \frac{\sigma_0 f S^{-1}\bar{X}}{\sqrt{f(\bar{X}^T S^{-1}\bar{X}) - \frac{c}{n}}}, \; if \frac{f(\mathbf{1}^T S^{-1}\bar{X})}{\sqrt{f(\bar{X}^T S^{-1}\bar{X}) - \frac{c}{n}}}\sigma_0 < 1 \\ \frac{S^{-1}\mathbf{1}}{\mathbf{1}^T S^{-1}\mathbf{1}} + f d_I[S^{-1}\bar{X} - \frac{\mathbf{1}^T S^{-1}\bar{X}}{\mathbf{1}^T S^{-1}\mathbf{1}}S^{-1}\mathbf{1}], otherwise \end{cases} \tag{6}$$

where $d_I = \sqrt{\frac{f(\mathbf{1}^T S^{-1}\mathbf{1})\sigma_0^2 - 1}{[f(\bar{X}^T S^{-1}\bar{X}) - \frac{c}{n}](f\mathbf{1}^T S^{-1}\mathbf{1}) - (f\mathbf{1}^T S^{-1}\bar{X})^2}}$ and $f = \frac{n-c-2}{n-1}$.

The correspondingly improved estimator for the optimal return can be estimated by $R_I = w_I^T \bar{X}$.

4 Empirical Experiments and Evaluation

In order to verify the performance of our approach, we conduct two experiments which explore different aspects of the problem. We compare our approach with two other approaches in the experiments: Approach 1: portfolios of 12 indexes produced by the improved estimation method illustrated in Section 3; Approach 2: portfolios of 12 indexes with equal weights. From the comparison with Approach 1 which does not consider trading strategies, we can test the usefulness of trading strategies, we can see to what extent our approach performs better than the movements of all indexes through the comparison with Approach 2.

In the first experiment, in order to test the performance of our approach in forecasting, the data is divided into four time periods. Each period consists of one year of data to estimate the optimal weights for corresponding portfolios.

The following month of data is used to compute the return of our approach and two other approaches. Thereafter, their performance is compared.

In the second experiment, we compare the performance of our approach with that of two other approaches in the whole period.

4.1 Data Description

In this paper, we illustrate the applicability of this trading strategy based portfolio selection approach on the investment of stock market indexes in 12 European Countries. As shown in Table 1, all these indexes are taken as price indices. We use daily closing prices of the component indexes listed in Table 1 from January 2006 to January 2012. The data is obtained from yahoo finance (http://finance.yahoo.com) in our experiments. Because different countries own different holidays and then lead to differences on trading days among the 12 countries, we delete the days that some countries are missing and only choose the trading days that all countries have trading. The method of computing the daily benefit for each index is listed in Section 3. We select two trading strategies from each of the four trading strategy classes for our application: Moving Average (MA(10,0.001), MA(50,0.005)), Filter Rules (FR(5,0.005), FR(10,0.001)), Support and Resistance (SR(5),SR(10)) and Stochastic Oscillator strategies (K/Dfast(7,7), K/Dslow(7,7)), the total number of trading strategies is eight.

Table 1. Trading Countries and Indexes

Country	Index	Country	Index	Country	Index
France	^FCHI	Austria	^ATX	Sweden	^OMXSPI
U.K.	^FTSE	Germany	^GDAXI	Portugal	GERAL.NX
Switzerland	^SSMI	Spain	^SMSI	Netherlands	^AEX
Norway	^OSEAX	Ireland	^ISEQ	Belgium	^BFX

4.2 Experimental Results

A. Testing for Forecasting

In this section the data is divided into four time periods for estimation and forecasting. These four periods are given in Table 2 [1]. Each period consists of one year of data for estimation. Here we use data beginning and ending at the middle of the year to avoid possible effect of the end of the calendar year. The benefit of the following month is then used to compute return of our approach (R_O), compared with that of Approach 1 (R_1) and Approach 2 (R_2).

In order to compute the optimal weights, we need to specify the maximum risk level σ_0 in each period. Here we use the standard deviation of the average of all the index benefit of each period to give us an approximate risk level. Suppose

[1] The number of the trading days of an index is too few from 09/09 to 12/09 and from 02/10 to 04/10, so here we delete the test from 07/09 to 06/10.

the standard deviation is $s(t), t = \{1, 2, 3, 4\}$, in this test we set $\sigma_0(t) = \frac{1}{5} \times s(t)$ to calculate the weights and then generate the corresponding return. The final results are shown in Table 2.

Table 2. The return of three approaches

Data	Test	R_O	R_1	R_2
07/06-06/07	07/07	-0.0028	-0.0361	-0.0345
07/07-06/08	07/08	-0.0352	-0.0229	-0.0432
07/08-06/09	07/09	0.1236	0.0325	0.0778
07/10-06/11	07/11	-0.0034	-0.0338	-0.0402
mean		0.0206	-0.0151	-0.0100

The means of the returns are displayed in the last row. From Table 2 we can see when compared with Approaches 1 and 2, our approach performs better. Specifically, our approach can generate more profit when financial markets go up and decrease loss when go down.

B. Backtesting

In this section, we mainly test the performance of our approach through the entire period from January 2006 to January 2012. There was a big change in financial markets during the period, namely financial crisis. The financial market was bull market before 2007 and investors can obtain profit easily, but the stock market experienced a big crash when encountered with the global financial crisis originated by the United States subprime mortgage market in 2007, after which the stock market was in oscillation. We want to see the performance of our approach under the fluctuant environment, when compared with two other approaches. The risk level is $\sigma_p = \frac{1}{5} \times s$, s here is the standard deviation of the average of all the index benefits of the entire period. The results of performance are shown in Fig.4.

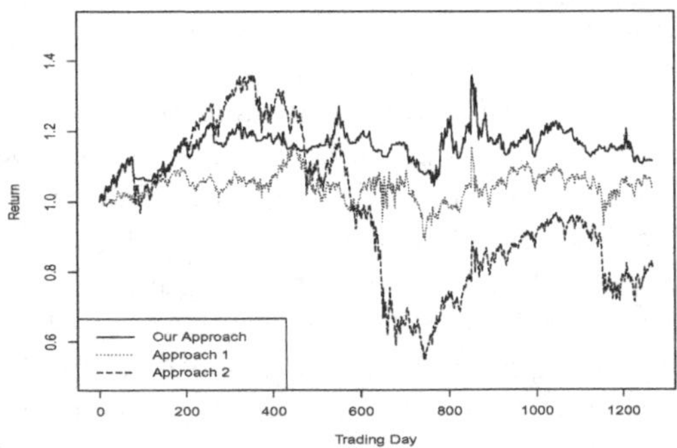

Fig. 4. The performance of three approaches

From Fig.4 we can see that our approach can help investors get a more stable return than other two approaches through the whole period. Investors who use our approach can avoid the big loss in the big crash created by the crisis. The results illustrate the practicality of our approach.

In addition, return volatility is an important factor that investors need to pay attention to. For securities, the higher the return volatility means a greater chance of a shortfall when selling the security in a future date, namely the investor need to face more risk. So here we compare the return volatility of our approach (risk level here is $\frac{1}{5}$) and two other approaches. The results are shown in Fig.5.

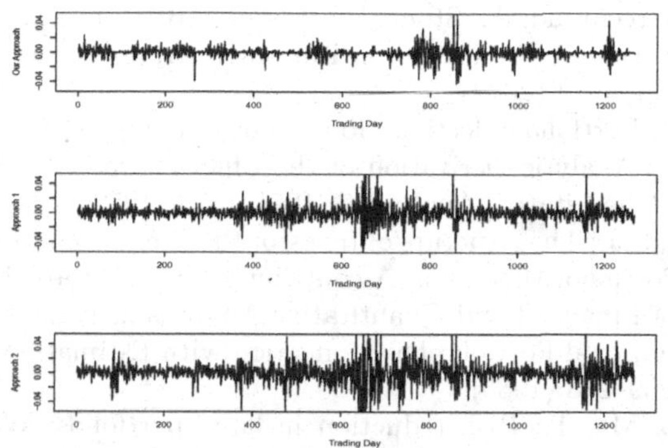

Fig. 5. The return volatility of three approaches

It is clear from Fig.5 that our approach own smaller return volatility than two other approaches. It means that when investors use our approach to make decision, the possibility of losing money is smaller.

A large amount of tests in index data of 12 countries have shown that our approach can lead to higher profit with lower risk when compared with two other approaches. From which we can see the usefulness of importing trading strategies and the superiority of our approach when compared with equally weighted indexes.

5 Conclusion

The marriage of trading strategy with portfolio is expected to greatly enhance the actionability of trading agents, while this has not been explored. This paper has proposed an approach for trading strategy based portfolio selection for trading agents by considering investor risk preference. This approach can enable trading agents to select optimal portfolios which best match investor's risk appetite based on recommended trading strategies. Six years of 12 individual markets data has been used for backtesting. The empirical results have indicated that our approach leads to effective increase of investment return for give risk, and thus strengthen the actionable capability of trading agents in the market.

References

1. Bai, Z.D., Liu, H.X., Wong, W.K.: Enhancement of the Applicability of Markowitz's Portfolio Optimization by Utilizing Random Matrix Theory. Mathematical Finance 19, 639–667 (2009)
2. Cao, L., Weiss, G., Yu, P.S.: A Brief Introduction to Agent Mining. Journal of Autonomous Agents and Multi-Agent Systems 25, 419–424 (2012)
3. Cao, L.B., Tony, H.: Develop Actionable Trading Agents. Knowledge and Information Systems 18(2), 183–198 (2009)
4. Cao, L.B., Luo, C., Zhang, C.Q.: Developing Actionable Trading Strategies for Trading Agents. In: International Conference on Intelligent Agent Technology, pp. 72–75 (2007)
5. Wellman, P.M., Greenwald, A., Stone, P., Wurman, R.P.: The 2001 Trading Agent Competition. In: Annual Conference on Innovative Application of Artificial Intelligence, vol. 13, pp. 4–12 (2002)
6. Markowitz, H.M.: Portfolio Selection. Journal of Finance 7, 77–91 (1952)
7. Merton, R.C.: An Analytic Derivation of the Efficient Portfolio Frontier. Journal of Financial and Quantitative Analysis 7, 1851–1872 (1972)
8. Cass, D., Stiglitz, E.J.: The structure of Investor Preferences and Asset Returns,and Separability in Portfolio Allocation: A contribution to the Pure Theory of Mutual Funds. Journal of Financial and Quantitative Analysis 2, 122–160 (1970)
9. Jorion, P.: International Portfolio Diversification with Estimation Risk. Journal of Business 58(3), 259–278 (1985)
10. Jagannathan, R., Ma, T.: Risk reduction in large portfolios: Why imposing the wrong constraints helps. Journal of Finance 58(4), 1651–1684 (2003)
11. Leung, P.L., Ng, H.Y., Wong, W.K.: An Improved Estimation to Make Markowitz's Portfolio Optimization Theory Users Friendly and Estimations Accurate with Application on the US Stock Market Inveatment. Working Paper (2011)
12. Elton, E.J., Gruber, M.J.: Modern portfolio theory, 1950 to due. Journal of Banking and Finance 21, 1743–1759 (1997)
13. Ioannis, A.V., Selman, B.: A principled study of the design tradeoffs for autonomous trading agents. In: Proc. of AAMAS 2003, pp. 473–480 (2003)
14. Sullivan, R., Timmermann, A.: Data-snooping, Technical Rule Performance, and the Bootstrap. Journal of Banking and Finance 54, 1647–1691 (1999)
15. Cao, L.B.: Data Mining and Multi-agent Integration (edited). Springer (2009)
16. Cao, L.B., Gorodetsky, V., Mitkas, P.A.: The Synergy of Agents and Data Mining. IEEE Intelligent Systems 24(3), 64–72 (2009)
17. Stock Trading Strategies, http://www.ashkon.com/predictor/strategies.html

On the Need of New Methods to Mine Electrodermal Activity in Emotion-Centered Studies

Rui Henriques[1], Ana Paiva[1,2], and Cláudia Antunes[1]

[1] DEI, Instituto Superior Técnico, Technical University of Lisbon, Portugal
[2] GAIPS, INESC–ID, Portugal
{rmch,ana.s.paiva,claudia.antunes}@ist.utl.pt

Abstract. Monitoring the electrodermal activity is increasingly accomplished in agent-based experimental settings as the skin is believed to be the only organ to react only to the sympathetic nervous system. This physiological signal has the potential to reveal paths that lead to excitement, attention, arousal and anxiety. However, electrodermal analysis has been driven by simple feature-extraction, instead of using expressive models that consider a more flexible behavior of the signal for improved emotion recognition. This paper proposes a novel approach centered on sequential patterns to classify the signal into a set of key emotional states. The approach combines SAX for pre-processing the signal and hidden Markov models. This approach was tested over a collected sample of signals using Affectiva-QSensor. An extensive human-to-human and human-to-robot experimental setting is under development for further validation and characterization of emotion-centered patterns.

1 Introduction

Wrist-worn biometric sensors can be used to track excitement, engagement and stress by measuring emotional arousal via skin conductance (SC), a form of electrodermal activity (EDA). Understanding EDA enables us to understand the role of the sympathetic nervous system in human emotions and cognition.

Although of critical value to neuroscience and psychophysiology, the study of EDA had been limited to the combined analysis of basic features: SC level, SC response amplitude, rate, rising time and recovery time. This method has a clear drawback – the discarding of flexible elicited behavior. For instance, a rising or recovering behavior may be described by specific motifs sensitive to sub-peaks or displaying a logarithmic decaying. This weak-differentiation among different stimuli response have led to poor emotional mappings, with EDA being mainly used just for the purpose of defining the intensity-axis of an emotional response.

This paper proposes a novel paradigm for the EDA analysis, the application of a sequence classifier over a symbolic approximation of the signal. This has the promise of disclosing emotions in real-time. In this way, scientific and clinical researchers can make dynamic adjustments to their protocols. Therapists

L. Cao et al.: ADMI 2012, LNAI 7607, pp. 203–215, 2013.

can gauge the effectiveness of in-session treatments. Professors can adapt their teaching strategies according to each students' response. Marketers can closely monitor focus-groups. Every person can use it to unfold unconscious behavior.

This paper is structured as follows. *Section 2* reviews the state-of-the-art on EDA and emotions theory in the context of biometric sensors. *Section 3* identifies the target problem. *Section 4* describes the proposed approach. *Section 5* proposes an experimental setting for the discovery of EDA emotion-driven patterns. *Section 6* identifies potential applications.

2 Related Research

This section provides a synthesized overview of the key contributions related to emotion recognition using physiological signals in general and EDA in particular.

2.1 Emotion Recognition Using Physiological Signals

Measuring physiological signals is increasingly necessary to derive accurate analysis from emotion-driven experiments. Physiological signals can surpass social masking and high context-sensitivity of image and audio analysis, track emotional changes that are less obvious to perceive, and provide complementary paths for their recognition (both cognitive and sensitive). However, their subtle, complex and subjective physical manifestation plus their idiosyncratic and variable expression within and among individuals present relevant key challenges.

The common problem in this context is to define a statistical learning method that can provide stable and successful emotional recognition performance. The main implication is to gain access to someone's feelings, which can provide important applications for human-computer interaction, conflict reduction, clinical research, well-being (augmented communication, self-awareness, therapy, relaxation) and education. Table 1 introduces a framework of five key questions to answer this problem. Good surveys with contributions gathered according to the majority of these axes include [15][38].

2.2 Emotions and the Electrodermal Activity

Electrodermal activity (EDA) is an electrical change[1] in the skin that varies with the activation of the *sympathetic* nervous system[2], which is responsible to activate positive excitement and anticipation, and to mobilize the body's fight-or-flight response by mediating the neuronal and hormonal stress response [1]. Electrical changes in the skin are a result of an increased emotional arousal or cognitive workload[3] that leads to an intense physical exertion, where brain

[1] The use of endosomatic methods is not target.

[2] Part of the autonomic nervous system responsible for the regulation of homeostatic mechanisms that require quick responses, complementary to "rest-and-digest" mechanisms triggered by the parasympathetic division.

[3] Involved neural pathways are numerous since excitatory and inhibitory influences on the sympathetic nervous system are distributed in various parts of the brain.

Table 1. The five decision-axes for recognizing emotions over physiological signals

Which physiological signals to measure?	Although EDA is the signal under analysis, its use can be complemented with other signals as, for instance, respiratory volume and rate if the goal is to recognize negative-valenced emotions, or heat contractile activity to distinguish among positve-valenced emotions [39]. Depending on the target emotions to assess, a combination of different modalities is desirable [15]. The key challenge is that modalities of emotion expression are broad (including electroencephalography; cardiovascular activity through electrocardiography, heart rate variability, cardiac output or blood pressure; respiratory activity; and muscular activity using electromyography), some yet being inaccessible or less studied (as blood chemistry, neurotransmitters and brain activity) and many others being too non-differentiated [29];
Which approach to follow?	User dependency, stimuli subjectivity and analysis time are the key axes [35][37]. In user-dependent approaches labeled EDA signals are vastly collected per user and the classification task for a target user is based on his historic pairs. User-independent approaches collect and use the pairs from a diversity of individuals to recognize emotions. Contrasting to "high-agreement" studies, in subjective experiments, the user is requested to self-report and/or to produce via mental imagery his response to a stimuli. Finally, the mining of a signal can be done statically or dynamically. This work targets the user-independent, non-subjective and dynamic evaluation quadrant;
Which models of emotions select?	The most applied models are the *discrete* model [13] centered on five-to-eight categories of emotions (there is considerable agreement in using happiness, sadness, surprise, anger, disgust, fear [15]) and the *dimensional* valence-arousal model [18] where emotions are described according to a pleasantness and intensity matrix. Other less commonly adopted models include the Ellsworth's dimensions and agency [27], Weiner's attributions and recent work (at MIT) focused on recognizing states that are a complex mix of emotions ("the state of finding annoying usability problems") [29];
Which experimental conditions to adopt?	The selected stimulus should evoke similar emotional reactions across individuals, be non-prone to contextual variations (time to neutralize the emotional state and to remove the stress associated with the experimental expectations), capture states of high and low arousal and valence to normalize the features, avoid multiple exposures (to not desensitize the subject), and provide reliable and reproducible methods according to existing guidelines [6][28]. The undertaken experiment is defined in section 5;
Which data processing and mining techniques to adopt?	Four steps are commonly adopted [15][6]. *First*, raw signals are pre-processes to remove contaminations (noise, external interferences and artefacts). Methodologies include segmentation; discard of initial and end signal bands; smoothing filters; low-pass filters such as Adaptive, Elliptic or Butterworth; baseline subtraction (to consider relative behavior); normalization; and discretization techniques [28][31][16][10]. *Second*, features are extracted. These features are statistical (mean, standard deviation), temporal (rise and recovery time), frequency-related and temporal-frequent (geometric analysis, multiscale sample entropy, sub-band spectra) [14]. The number may vary between a dozen to hundreds of features depending on the number and type of the adopted signals [15]. Methodos include rectangular tonic-phasic windows; moving and sliding features (as moving and sliding mean and median); transformations (Fourier, wavelet, empirical, Hilbert, singular-spectrum); principal, independent and linear component analysis; projection pursuit; nonlinear auto-associative networks; multidimensional scaling; and self-organizing maps [14][19][15]. *Third*, features that might not have significant correlation with the emotion under assessment are removed. This increases the classifiers' performance by reducing noise, enabling better space separation, and improving time and memory efficiency. Methods include: sequential forward/backward selection, sequential floating search, "plus t-take-away r" selection, branch-and-bound search, best individual features, principal component analysis, Fisher projection, classifiers (as decision tress, random forests, bayesian networks), Davies-Bouldin index, and analysis of variance methods [15][6]. *Finally*, a classifier is learned using the previously selected features. Methods include a wide-variety of deterministic and probabilistic classifiers, with the most common including: k-nearest neighbours, regression trees, random forests, Bayesian networks, support vector machines, canonical correlation analysis, neural networks, linear discriminant analysis, and Marquardt-back propagation [24][15][25].

stimulus may lead to sweating[4]. The skin is believed to be the only organ to react only to the sympathetic part of the nervous system, allowing its measurements to get a more accurate reading [8].

By monitoring EDA is possible to detect periods of excitement, stress, interest and attention. However, heightened skin conductance is also related with engagement, hurting, intrigue, distress and anticipation ("the unknown behind the wall") [1]. In fact, EDA is influenced primarily by the activation of an inhibition function that is involved in responding to punishment, passive avoidance or frustrative non-reward, which are different forms of anxiety [8]. These recent clarifications on the role of EDA responses, require careful experimental conditions and, as target by this paper, more robust methods for their mining.

On one hand, measuring EDA has clear advantages: sympathetic-centered response, neuro-anatomical simplicity, trial-by-trial visibility, utility as a general arousal and attention indicator, significance of individual differences (reliably associated with psychopathological states), and its simple discrimination after a single presentation of a stimulus. On the other hand, EDA has a relatively slow-moving response (latency of the elicited response and tonic shifts between 1 and 3s and varying among individuals [8]), requires lengthy warm-up periods, and has multiple influences that may be either related with the subject attention and personal significance, stimuli activation, and affective intensity.

The variety of electrodermal phenomena can be understood by mining changes in tonic SC level (SCL) and phasic SC response (SCR), related to tonic or phasic sympathetic activation. Researchers have found that tonic EDA is useful to investigate general states of arousal and alertness, while phasic EDA is useful to study multifaceted attentional processes (related to novelty, intensity, and significance), as well as individual differences in both the normal and abnormal spectrum [8]. Although these are important achievements, there is still the need to verify if, under controlled experimental conditions, the inclusion of advanced signal behavior can increase or not the accuracy of a target classifier.

Experimental Evidence. Historical EDA studies had been focused on learning efficiency, response speed and, as target by this paper, emotional appraisal. Three distinct types of experiments have been done.

First: experiments using discrete stimuli. Experiments with brief and isolated stimuli, include the study of: *innocence* using the guilty knowledge test [22]; *familiarity* by distinguishing between meaningful and unfamiliar stimuli [1]; *relevance* through non-balanced occurrence of a stimuli category or through elicitation of priorities [2]; *affective valence* (although not good in discriminating along the positive-negative axis, EDA was, for instance, found to be higher for erotic pictures or striking snakes than for beautiful flowers or tombstones [17]); and *planning and decision-making* processes via the "somatic marker" hypothesis [34]. Backward masking is often used to prevent awareness of conditioning stimulus by preventing its conscious recognition [8]. The great challenge when

[4] EDA has both a functioning role (maintain body warmth, and priming the body for action) and evolutionary meaning (protection from grasping injury).

recognizing emotion is that the elicit response are considered to be part of the orienting response to novel stimuli, which influence should be removed.

Second: experiments using continuous stimuli. When studying effects of long-lasting stimuli, SCL and frequency of spontaneous SCRs (NS-SCRs) are key measures. Experiments include the study of: *strong emotions* reproducing, for instance, genuine states of fear (highest SCLs) and anger (greatest NS-SCRs) [1]; *reappraisal* through authentic, forbidden and awarded emotional display; physical and mental *performance*; *attention* (affecting rising and recovery time in vigilance tasks); and different forms of *social interaction* involving, for instance, *judgment* (NS-SCRs rate inversely related to the judged permissiveness of a questioner), *distress invocation* through the study of relationships, or the *contagious effect* by relating, for instance, heightened autonomic arousal with living with over-involved individuals [8][7]. Energy mobilization seems to be the driver for tasks that either require an effortful allocation of attentional resources or, but not necessarily exclusive, invoke the concepts of stress and affect.

Third: potential long-term experiments targeting personal traits. High NS-SCRs rate and slow SCR habituation are used to define a trait called lability with specific psychophysiological variables [11]. Traits have been defined according to: *information processing* [32], *operational performance*, *brain-side activation* through studies with epileptic individuals or recurring to electrical stimulation (right-side of limbic structures stimulation increases more SCR than the left) [1], *sleeping patterns* [21], *age* [8], *psychopathology* (mainly diagnosable schizophrenia and subjects with tendency to emotional withdrawal and conceptual disorganization, with different traits regarding to the *SCR conditioning* (revealing paths to emotional detachment as absence of remorse and antisocial behavior as pathological lying and substance abuse), *tonic arousal*, and *response to mild innocuous tones* [23]. These results suggest that hypo- or hyper-reactivity to the environment may interfere with fragile cognitive processing in ways that underline vulnerabilities in the areas of social competence and coping.

Approaches to Analyze EDA. Current approaches are focused on features' extraction from the signal, neglecting its motifs. When measuring EDA from discrete stimuli, the key adopted feature is the SCR amplitude. The response latency, rise time and half recovery time are sporadically adopted, although their relation to psychophysiological processes remain yet unclear.

When studying prolonged stimulation, both specific and spontaneous responses are considered. NS-SCRs frequency is the feature of interest, which can be easily computed using a minimum amplitude as threshold. An alternative is to compute the SCL, which can either include or exclude the specific responses periods depending on the experimental conditions (continuous or sporadic stimuli presentation). For the latter case, a latency window criterion is required.

Finally, the analysis of traits also recurs to NS-SCRs rate, SCL, response amplitude and habituation. The challenge is on whether to use or not a range correction, by capturing the maximum and minimum EDA values during a session. Both relative and absolute approaches can be found in the literature, with pros-and-cons [8] and alternatives [3]. Test-retest reliability, psychometric principles and questionnaires are crucial to view EDA response as a trait [8].

3 The Problem

The target problem of this work is to generalize the EDA sequential behavior for different emotions and to assess their effect in emotion classification accuracy. In particular, for emotions elicited in human-robot vs. human-human interaction. Emotion-driven EDA behavior can be learned by statistical models to dynamically classify emotions. The contribution of this work is on proposing a novel approach for this problem centered on an expressive pre-processing step followed by the direct application of a sequence classifier, instead of performing traditional methods of feature-driven analysis. Initial evidence for the good performance of this approach over a preliminary collection is presented.

4 The Proposed Approach

This section proposes a new paradigm for emotion-recognition from EDA, collapsing the traditional four-step process into a simpler and more flexible two-stage process. There are two core strategies: to rely upon a good representation of the signal, and to mine sequential patterns instead of retrieving domain features.

4.1 Experimental Conditions and Data Properties

The collected EDA signals were obtained using wrist-worn Affectiva-QSensors[5] and closely-controlled experimental procedures[6]. The wireless connectivity of the adopted sensors enables the real-time classification of emotional states.

Additionally, the following signals were collected using Affectiva technology: facial expression series, skin temperature and three-directional motion. Although this paper is centered on the analysis of EDA, the joint analysis of the adopted signals (multivariate time series mining) is of additional interest and can be done by extending the dynamic Bayesian networks. Currently, the last two signals are being currently used to affect the EDA signal: skin temperature to weight SCL by correcting the individual reaction to room temperature, and body intense movements to smooth correlated EDA variations. Facial recognition is adopted just for post-experimental validation and interpretation.

[5] Data captured is considered as reliable as the tethered system developed by BIOPAC, often used in physiological research, and is currently being adopted over a hundred universities, which enable a standardized way of comparing experiments.

[6] Includes measuring of very high and low states of EDA, conservative signal stabilization criteria, standardized stimuli presentation and context reproducibility.

4.2 Processing the Signal Using SAX

Since our goal is sequential data classification, we are interested in one approach that simultaneous supports: *i)* reduced dimensionality and numerosity, and *ii)* lower-bounding by transforming real-valued EDA into a symbolic representation. First, reducing the high dimensionality and numerosity of signals is critical because all non-trivial data mining and indexing algorithms degrade exponentially with dimensionality. Second, symbolic representations allow for the application of more expressive techniques like hidden Markov models and suffix trees.

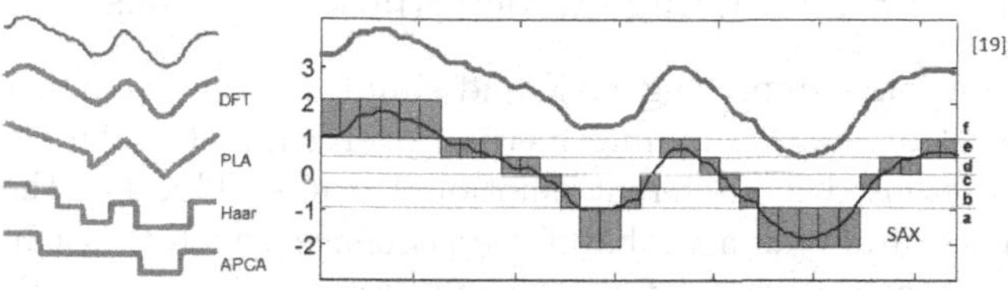

Fig. 1. Commonly adopted representations vs. SAX [20]

While many symbolic representations of time series exist, they suffer from two critical flaws: *i)* mining algorithms scale poorly as the dimensionality of the symbolic representation is not changed, and *ii)* most of these approaches require one to have access to all the data, before creating the symbolic representation. This last feature explicitly thwarts efforts to use the representations with streaming methods, required for a dynamic emotion recognition from EDA. To the best of our knowledge only Symbolic ApproXimation (SAX) provides a method for performing operations in the symbolic space while providing the lower bounding guarantee [20][33]. SAX is demonstrated to be as competitive or superior as alternative representations for time series classification. SAX allows a time series of arbitrary length n to be reduced to a string of w-length ($w<n$, typically $w \ll n$) with dimension or alphabet size $d>2 \in \mathbb{N}$. To mine signals in main memory for real-time classification purpose, w and d parameters have to be carefully chosen. Fig.1 compares SAX with the four most common adopted alternatives [12]. A raw time series of length 128 is transformed into the word ffffffeeeddcbaabceedcbaaaaacddee.

This work implemented SAX in two steps. Firstly, the signal is transformed into a Piecewise Aggregate Approximated (PAA) representation, which provide a well-documented method to reduce dimensionality. Secondly, the PAA signal is symbolized into a discrete string allowing lower bounding, which can be useful to perform distance metrics for recognizing emotion-centered patterns needed for the classification task. Here, an important technique with visible impact in accuracy, is the use of a Gaussian distribution over the normalized signals to produce symbols with equiprobability recurring to statistical breakpoints [20]. In Fig.1, breakpoints define the boundary criteria across symbols.

Different criteria may be adopted to fix the signal dimensionality and numerosity values. The normalization step can be done with respect to all stimulus, to a target stimuli, to all subjects and to the available responses of a target subject. Additionally, two mutually-exclusive strategies can be defined to deal with variable signal numerosity. First, a ratio to reduce numerosity can be uniformly applied across the collected signals. Second, piecewise aggregation can be adopted to balance the signals numerosity with respect to a particular emotion label. Note that since the temporal axis is no longer absolute, relevant information is lost and, therefore, this second strategy should only be adopted when complemented by the first.

4.3 Mining the Signal Using Hidden Markov Models

After a pre-processing step using SAX and simple artefact-removal techniques, there is the need to apply a mining method over sequential data to categorize the behavior presented for different emotional states. This has the promise of increasing the accuracy as a probabilistic sequence generation can be learned from the raw processed signal for each emotion-class, instead of loosing key behavioral data through the computation of a simple set of metrics. Some of the most popular techniques include recurrent neural networks, dynamic Bayes networks and adapted prototype-extraction [5][26].

This paper proposes the use of hidden Markov models (HMM), a specific type of a dynamic Bayes network, due to their stability, documented performance in healthcare domains, simplicity and flexible parameter-control [30]. In particular, [4] proposes that, at least within the paradigm offered by statistical pattern classification, there is no general theoretical limit to HMMs performance given enough hidden states, rich enough observation distributions, sufficient training data, adequate computation, and appropriate training methods.

Markov models simplicity derives from the assumption that future predictions are independent of all but the most recent observations. In a HMM, an underlying and hidden automaton of discrete states follows a Markov constraint and the probability distribution of the observed signal state at any time is determined only by the current hidden state. Given a set of training signals labeled with a specific emotion, the core task is to learn the transition and generation probabilities of the hidden automaton per emotion. This is done in practice by maximizing the likelihood function iterations of an efficient forward-backward algorithm until the transition and generation probabilities converge [30][5]. Finally, given a non-labeled signal, the selection of the emotion can be naively classified by evaluating the generation probability of the exponential paths generated from each learned automaton lattices, and by selecting the path having the highest probability. For this purpose, the Viterbi algorithm was selected [36].

In order to define the input parameters for the HMMs two strategies may be consider. First, a sensitivity-analysis over the training instances per emotion to maximize accuracy. Second, parameter-definition based on the signal properties (e.g. high numerosity leads to an increased number of hidden states). Additionally, an extension to traditional HMMs can be made to deal with multiple

EDA signals with varying dimensionality (smoothed and pronounced EDA). This aims to increase the accuracy of the target approach by providing multiple paths to select the label, since one path may not be the best for two different emotions. Currently, this is done by computing the joint probability of the different paths.

5 Validation of Our Approach

The proposed EDA mining approach were applied over a small set of subjects with preliminary but interesting conclusions. This section reviews them and characterizes the experimental setting to be adopted for further validation and emotion-driven EDA characterization.

5.1 Preliminary Results

Preliminary evidence of the utility is described in Table 2. Due to the small sized of the collected sample of stimuli-response EDA, no quantitative analysis on the accuracy, specificity and sensitivity classification metrics is provided.

5.2 Next Steps

To gain further insight of the EDA response pattern to specific emotion-oriented stimuli, we are undertaking a tightly-controlled lab-experiment. We expect to have around fifty subjects and, at least, forty skin-conductive subjects with valid collections. Since eight different stimuli (five emotion-centered and two others) will be used per subject, our final dataset will have above three hundred collected signals, with each stimuli having above thirty instantiations, which satisfies the statistical requirements of hidden Markov models.

Additionally, facial recognition, skin temperature, body 3-dimensional motion and video-audio recording will be captured. A survey will be used to categorize individuals according to the Myers-Briggs type indicator and for a complementary context-dependent analysis of the results. The target emotions are empathy, expectation, positive-surprise (unexpected attribution of a significant incremental reward), stress (impossible riddle to solve in a short time to maintain the incremental reward) and frustration (self-responsible loss of the initial and incremental rewards dictated by the agent). The adopted reward for all subjects is one cinema-session ticket-offer. The stimulus will be presented in the same order in every experience and significant time will be provided between two stimulus to minimize influence, although noise propagation across stimuli is a necessary condition in multiple-stimulus experiment. Equivalent scenarios will be used for human-human and human-robot interaction, with subjects being randomly selected to attend one scenario. The robots used for this experience will be EMYS and NAO[7].

[7] `http://emys.lirec.ict.pwr.wroc.pl` and http://www.aldebaran-robotics.com

Table 2. Initial observations of the target approach over EDA samples

Challenge	Observations
Expressive behavior	An intricate observation was the sensitivity of the learned HMMs to expressive behavior as peak-sustaining values (e.g. as a response to warm hugs) and fluctuations (e.g. for elicited anger). Such behavior is hardly measured by feature-extraction methods since they loose substantial amounts of potential relevant information during the computation process and are strongly dependent on directive thresholds (e.g. peak amplitude to compute frequency measures). We expect that, with the increase of available signals, HMMs are able to learn internal transitions that capture smoothed shapes per emotion, which enable the discrimination of different types of rising and recovering responses following sequential patterns with flexible displays (e.g. exponential, "stairs"-appearance). The number of discrete hidden values is an important variable for this expressivity. In our current implementation, a sensitivity-analysis over the training instances is performed per emotion until maximum accuracy is achieved;
Numerosity differences	Two strategies were adopted to overcome this challenge. First, signals as-is (with their different numerosity) were given as input to HMMs as dynamic Bayesian networks are able deal with this aspect (note, for instance, the robustness of HMM on detecting hand-writing text with different sizes in [5]). Second, the use of piecewise aggregation analysis by SAX can be used to normalize different signals with respect to their numerosity. However, since a good piece of temporal information is lost (as latency, rising and recovery time), this temporal normalization is performed per stimuli with respect to the average length of responses. This second strategy increases significantly the performance of HMMs if the following algorithmic-adaptation is performed: the input signal is mapped into the standard-numerosity of each emotion in order to assess the probability of being generated by each emotion-centered Markov model;
SCL differences	One of the key challenges is to deal with individual differences in terms of SCL and SCRs amplitude under the same emotion. The normalization step in traditional approaches fails to answer this challenge as SCL and response amplitude are not significantly correlated (e.g. high SCL does not mean heightened SC responses). The Gaussian distribution for dimensionality control used by SAX provide a simple method to smooth this problem. Additionally, our implementation supports both absolute and relative criteria to mine EDA signals, with the scaling strategy being done with respect to all stimulus, to the target stimuli, to all subjects or to subject-specific responses;
Lengthy responses	Rising and habituation time provide a poor framework to study lengthy responses as, for instance, response to astounding stimulus (where spontaneous amplitude-varying relapses are present). This expressive behavior can still be considered in lengthy series by increasing the number of hidden states of the target HMM. Our implementation enables a dynamic adaptation of HMM parameters based on the average length of response to each stimuli;
Peak sensitivity	Our approach has the promise of overcoming the limitations of feature-based methods when dealing with fluctuations of varying amplitude and temporal distance (for instance, non-periodic relapses). This is done by controlling dimensionality using SAX. A range of values for dimensionality can be adapted, with two main criteria being adopted to increase the accuracy of our classifiers: mapping the raw signals into low-dimensional signals to capture smoothed behavior (e.g. alphabet size less than 8) and into high-dimensional signals to capture more delineated behavior (e.g. alphabet size above 10). Currently, two HMM are being generated for each of the strategy, with the joint classification probability being computed to label a response. However, in future work, it is expected an adaptation of the adopted HMMs to deal with multiplicity of signals, each one embedding different dimensional criteria;

6 Applications

The main implication of the potential gains in accuracy for recognizing emotions is an improved access to someone's feelings. One key area covered by the recent efforts to integrate data mining and agent interaction [9]. In the target experiment, this has direct application in *human-robot interaction*. Additional applications include: *clinical research* (emotion-centered understanding of addiction, affect dysregulation, alcoholism, anxiety, autism, attention deficit hyper- and hypoactivity, depression, drug reaction, epilepsy, menopause, locked-in syndrome, pain management, phobias and desensitization therapy, psychiatric counseling, schizophrenia, sleep disorders, and sociopathy); *well being* as the study of the effect of relaxation techniques like breathing and meditation; *marketing* to understand the emotions evoked by a message; *conflict reduction* in schools and prisons by early detection of hampering behavior (particularly important with autistics who have particular trouble understanding theirs and others feelings); *education* through the use of real-time emotion-centered feedback from students to escalate behavior and increase motivation; and many others as biofeedback, EDA-responsive games and self-awareness enhancement.

7 Conclusion

This work introduces a novel paradigm to analyze electrodermal activity in emotion-centered experiments. For this purpose, it proposes an approach centered on an expressive pre-processing step using SAX followed by the application of hidden Markov models over the processed sequential data. This has the benefit of overcoming the limitations of traditional methods based on feature extraction, namely limitations to deal with expressive behavior (flexible relation of both temporal and amplitude axis through patterns) and with individual response differences related to signal dimensionality and numerosity. Multiple criteria for modeling the signal and for defining the classifier parameters are proposed, with the labeling step relying on the calculus of joint probabilities.

These results were supported by initial observations from a collected sample of signals. An extended experimental study is being undertaken for further validation and to characterize the differences among emotion-driven EDA patterns.

Acknowledgment. This work is partially supported by *Fundação para a Ciência e Tecnologia* under the PhD grant SFRH/BD/75924/2011 and the research project D2PM (PTDC/EIA-EIA/110074/2009).

References

1. Andreassi, J.: Psychophysiology: Human Behavior and Physiological Response. In: Psychophysiology: Human Behavior & Phy. Response. Lawrence Erlbaum (2007)
2. Ben-Shakhar, G.: A Further Study of the Dichotomization Theory in Detection of Information. Psychophysiology 14, 408–413 (1977)

3. Ben-Shakhar, G.: Standardization within individuals: A simple method to neutralize individual differences in skin conductance. Psychophy 22(3), 292–299 (1985)
4. Bilmes, J.A.: What hmms can do. IEICE Journal E89-D(3), 869–891 (2006)
5. Bishop, C.M.: Pattern Recognition and Machine Learning (Information Science and Statistics). Springer-Verlag New York, Inc., Secaucus (2006)
6. Bos, D.O.: Eeg-based emotion recognition the influence of visual and auditory stimuli. Emotion 57(7), 1798–1806 (2006)
7. Brown, G., Birley, J., Wing, J.: Influence of family life on the course of schizophrenic disorders: a replication. B.J. of Psychiatry 121(562), 241–258 (1972)
8. Cacioppo, J., Tassinary, L., Berntson, G.: Handbook of psychophysiology. Cambridge University Press (2007)
9. Cao, L.: Data mining and multi-agent integration. Springer, Dordrecht (u.a) (2009)
10. Chang, C., Zheng, J., Wang, C.: Based on support vector regression for emotion recognition using physiological signals. In: IJCNN, pp. 1–7 (2010)
11. Crider, A.: Electrodermal response lability-stability: Individual difference correlates. In: Prog. in Electrod. Research, vol. 249, pp. 173–186. Springer, US (1993)
12. Ding, H., Trajcevski, G., Scheuermann, P., Wang, X., Keogh, E.J.: Querying and mining of time series data: experimental comparison of representations and distance measures. PVLDB 1(2), 1542–1552 (2008)
13. Ekman, P., Friesen, W.: Universals and cultural differences in the judgments of facial expressions of emotion. J. of Personality and Social Psychology 53, 712–717 (1988)
14. Haag, A., Goronzy, S., Schaich, P., Williams, J.: Emotion Recognition Using Biosensors: First Steps towards an Automatic System. In: André, E., Dybkjær, L., Minker, W., Heisterkamp, P. (eds.) ADS 2004. LNCS (LNAI), vol. 3068, pp. 36–48. Springer, Heidelberg (2004)
15. Jerritta, S., Murugappan, M., Nagarajan, R., Wan, K.: Physiological signals based human emotion recognition: a review. In: 2011 IEEE 7th International Colloquium on Signal Processing and its Applications (CSPA), pp. 410–415 (2011)
16. Katsis, C., Katertsidis, N., Ganiatsas, G., Fotiadis, D.: Toward emotion recognition in car-racing drivers: A biosignal processing approach. IEEE Transactions on Systems, Man and Cybernetics, Systems and Humans 38(3), 502–512 (2008)
17. Lang, P.J., Bradley, M.M., Cuthbert, B.N.: International affective picture system (IAPS): Technical Manual and Affective Ratings. NIMH (1997)
18. Lang, P.: The emotion probe: Studies of motivation and attention. American Psychologist 50, 372–372 (1995)
19. Lessard, C.S.: Signal Processing of Random Physiological Signals. Synthesis Lectures on Biomedical Engineering, Morgan and Claypool Publishers (2006)
20. Lin, J., Keogh, E., Lonardi, S., Chiu, B.: A symbolic representation of time series, with implications for streaming algorithms. In: ACM SIGMOD Workshop on DMKD, pp. 2–11. ACM, New York (2003)
21. Lorber, M.F.: Psychophysiology of aggression, psychopathy, and conduct problems: a meta-analysis. Psychological Bulletin 130(4), 531–552 (2004)
22. Lykken, D.T.: The gsr in the detection of guilt. J. A. Psych. 43(6), 385–388 (1959)
23. Lykken, D.: A study of anxiety in the sociopathic personality. U. Minnesota (1955)
24. Maaoui, C., Pruski, A., Abdat, F.: Emotion recognition for human-machine communication. In: IROS, pp. 1210–1215. IEEE/RSJ (September 2008)
25. Mitsa, T.: Temporal Data Mining. In: DMKD. Chapman & Hall/CRC (2009)
26. Murphy, K.: Dynamic Bayesian Networks: Representation, Inference and Learning. Ph.D. thesis, UC Berkeley, Computer Science Division (July 2002)

27. Oatley, K., Keltner, Jenkins: Understanding Emotions. Blackwell P. (2006)
28. Petrantonakis, P.C., Hadjileontiadis, L.J.: Emotion recognition from eeg using higher order crossings. Trans. Info. Tech. Biomed. 14(2), 186–197 (2010)
29. Picard, R.W.: Affective computing: challenges. International Journal of Human-Computer Studies 59(1-2), 55–64 (2003)
30. Rabiner, L., Juang, B.: An introduction to hidden Markov models. ASSP Magazine 3(1), 4–16 (2003)
31. Rigas, G., Katsis, C.D., Ganiatsas, G., Fotiadis, D.I.: A User Independent, Biosignal Based, Emotion Recognition Method. In: Conati, C., McCoy, K., Paliouras, G. (eds.) UM 2007. LNCS (LNAI), vol. 4511, pp. 314–318. Springer, Heidelberg (2007)
32. Schell, A.M., Dawson, M.E., Filion, D.L.: Psychophysiological correlates of electrodermal lability. Psychophysiology 25(6), 619–632 (1988)
33. Shieh, J., Keogh, E.: isax: indexing and mining terabyte sized time series. In: ACM SIGKDD, KDD 2008, pp. 623–631. ACM, New York (2008)
34. Tranel, D., Damasio, H.: Neuroanatomical correlates of electrodermal skin conductance responses. Psychophysiology 31(5), 427–438 (1994)
35. Villon, O., Lisetti, C.: Toward recognizing individual's subjective emotion from physiological signals in practical application. In: Computer-Based Medical Systems, pp. 357–362 (2007)
36. Viterbi, A.: Error bounds for convolutional codes and an asymptotically optimum decoding algorithm. IEEE Trans. on Inf. Theory 13(2), 260–269 (1967)
37. Vyzas, E.: Recognition of Emotional and Cognitive States Using Physiological Data. Master's thesis. MIT (1999)
38. Wagner, J., Kim, J., Andre, E.: From physiological signals to emotions: Implementing and comparing selected methods for feature extraction and classification. In: ICME, pp. 940–943. IEEE (2005)
39. Wu, C.K., Chung, P.C., Wang, C.J.: Extracting coherent emotion elicited segments from physiological signals. In: WACI, pp. 1–6. IEEE (2011)

Author Index